ALSO BY ALAN WALKER

The Nariokotome Homo erectus *Skeleton* (with Richard Leakey)
The Human Skeleton (with Pat Shipman and David Bichell)
Prosimian Biology (with R. D. Martin and G. A. Doyle)

ALSO BY PAT SHIPMAN

Life History of a Fossil
The Human Skeleton (with Alan Walker and David Bichell)
The Neandertals (with Erik Trinkaus)
The Evolution of Racism

THE
WISDOM
OF THE
BONES

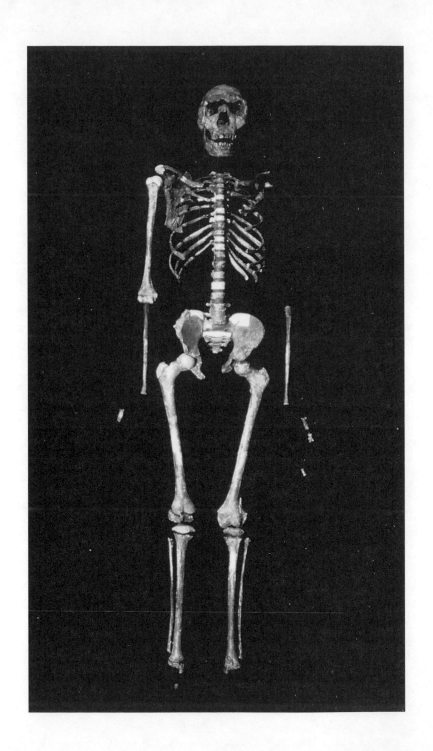

THE
WISDOM
OF THE
BONES

IN SEARCH OF HUMAN ORIGINS

ALAN WALKER

AND

PAT SHIPMAN

ALFRED A. KNOPF   NEW YORK   1996

THIS IS A BORZOI BOOK
PUBLISHED BY ALFRED A. KNOPF, INC.

Library of Congress Cataloging-in-Publication Data
Walker, Alan.
The wisdom of bones : in search of human origins / Alan Walker and
Pat Shipman. — 1st ed.
p.   cm.
Includes bibliographical references and index.
ISBN 0-679-42624-8
1. Homo erectus—Kenya.   I. Shipman, Pat.   II. Title.
GN284.W35   1996
573.3—dc20   95-37525
CIP

Manufactured in the United States of America
First Edition

*For Kamoya Kimeu and
the hominid gang*

CONTENTS

Eight pages of photographs follow page 116.

ACKNOWLEDGMENTS

Many people and organizations contributed to the success of the research described here—and to the completion of this book. First and foremost, I owe a tremendous debt to Kamoya Kimeu and the gifted members of the hominid gang, which has included, at various times, Christopher Kiarie, Musa Kyeva, Benson Kyongo, Wambua Mangao, Mwongela Muoka, Joseph Mutaba, Peter Nzube Mutiwa, and Bernard Ngeneo. In camp and on the road, the team and I were helped by Daniel Mutinda, Omari Hassan, Solomon son of Derekich, and Kabai Nd'unda. Meave and Richard Leakey were partners in this, as in so many projects, and only they can know how grateful I am. Without these people, none of this would have been possible.

The National Museums of Kenya and the National Geographic Society provided vital financial support for the fieldwork. Much of the research and analysis was conducted with help from the John Simon Guggenheim Memorial Foundation and a timely and ever appreciated fellowship from the John D. and Catherine T. MacArthur Foundation. The Wenner-Gren Foundation gave a grant to my wife and coauthor for a study of Eugène Dubois, the results of which we have used here. In Leiden, John de Vos shared information about Eugène Dubois and provided open access to the invaluable Dubois Archives, while Paul Storm provided translations from the Dutch.

This is a personal account of my adventures with *Homo erectus*, not a comprehensive scientific treatment. For this reason, the colleagues—living and dead—who have studied *Homo erectus* and influenced my thinking are too numerous to mention specifically here or in the text, although their conclusions and ideas have been folded into the texture of this story. I hope they and those who collaborated directly in finding and analyzing 15K understand my gratitude and respect for their contributions. In addition, D. C. Johanson, Ann MacLarnon, G. W. H. Schepers, B. Holly Smith, C. Fred Spoor, Carl Swisher III, Phillip V. Tobias, and Tim White helped by granting interviews or answering queries by mail. Gail Lewin and Chris Wadman read the manuscript and made cogent suggestions to improve it. I thank them all.

Books are not produced by authors alone, and for their assistance in bringing this story to the public I thank Jonathan Segal and his able assistant, Ida Giragossian, of Alfred A. Knopf.

Readers may wonder why this book is written in the first person, yet my wife is listed as a coauthor. It is my story, but she wrote most of the words.

A.W.

THE
WISDOM
OF THE
BONES

PROLOGUE

I really shouldn't be here.

A Turkana woman has just walked from behind a mass of hanging doum palm leaves into the shade in which I am sitting. With each step, her long, brown goat-skin skirt kicks up puffs of dust behind her. Above, she wears only a mass of blue and white beads; they pile up under her chin and extend out onto her thin shoulders. Her smooth, small head is shaved, for beauty, except for a straggly, midline topknot.

Suddenly she sees me and stops absolutely dead, as if she has had a stroke in midstride. Then she shuts her eyes. I know exactly what she is doing: she is going to wait a few seconds before she looks my way again. She expects—she *hopes*—that I will have gone by then, that I will prove to have been a hallucination. She has never before seen a long, sunburned Englishman in tennis shorts and a loose-fitting, sweat-soaked shirt, sitting under a doum palm.

I don't hesitate. While her eyes are still closed, I get up quickly and move down the bank into the furnace blast of light and heat. I walk quickly down the sand river. This is her place, after all. Where I grew up, the land was always green and wet and the rivers were full to overflowing. Where she lives, here on the west side of Lake Turkana in northern Kenya, the land is dry and buff colored and all the plants have thorns on them. In some years, there is no rain. Her river is nothing like my own river Soar, in Leicestershire, home of roach and bream. The Soar is a glistening flow that moves slowly through meadows filled with contented, fat cows. Her river is a double strand of acacias or palms, separated by an area that is sandier and flatter than most: a good place to drive, a sort of bush highway. And if you dig there, you'll find water—if you know where to dig (the trees are the clue; they sink their roots into the damp hidden deep below) and if you dig deep enough.

When I am far enough downstream not to disturb her I find another patch of shade and consider my position. What am I doing here? More to the point, what are *we* doing here: not me and the woman, but me, Mac, Richard, and the others. I am here only because there is absolutely no point in looking for early human ancestors—the missing links between apes and humans—at home. They didn't live there; they lived here. Just back from the river there is a system of gorges and gullies that cut into cream, red, and tangerine sediments nearly four million years old. It is a good age, a relatively unknown age in human evolution, and these arid badlands usually pull me like a magnet, because I need to know where we all came from; I need to know, in a much bigger sense, what all of us are doing here. Mac, Richard, and the others feel the same, I think: finding missing links will lead to the answers. It is that simple.

The Turkana woman and I are separated by nearly all our experiences of daily life and by the differences in culture through which we interpret those experiences. We have no language in common and few shared concerns. All we share, seemingly, is our familiarity with this desolate patch of desert. But I know there is something more, something greater and more fundamental that links us to each other as irrevocably as spots cling to the leopard, and that is our humanity.

What is it to be human? She and I probably have different ideas, different mythologies of how humans were created, tales told around our respective campfires for many years. Storytelling is perhaps an integral human trait, but the content of stories is more often emotional than literal truth. What we have in common, this unknown woman and I, is far deeper and more powerful than any set of beliefs. It is a way of being, a set of bones and muscles and nerves that work in a certain way, making it easy to walk upright or to hold and manipulate an object, yet difficult to slice through raw meat or hide with our teeth. It is a brain constructed with interconnections here and none there, with room for enormous learning and change, with a large capacity for memory, with an implacable need for social interactions. Biologically, evolutionarily, all this is dictated; "humans are this type of animal and not that" is written in our genes. But we do not know the details of the *this* and *that*; their meaning is a question we can pose only to the fossil record that documents our evolutionary development. Finding missing links—discovering the fossils and filling in the gaps in our evolutionary story—is the only way to determine the answers. There are few bigger questions in the world than "who are we?" and the chance of finding an answer is worth the sweaty discomfort, the loneliness of being away from home and family, the occasional genuine danger, and the frequent disappointment.

I should be looking in the gullies now, where erosion is more likely to have exposed whatever fossils are there, but the heat and unusual humidity have got to me today. Inside the gullies, it is 135°F. Usually it is dry heat, like a blast furnace: bearable but posing a real risk of dehydration. Today, peculiarly, it is humid and that is worse. There is no air, no trace of a breeze, and I notice that I tend to walk higher on the slopes so that I can catch a draft. It may seem I am out of my element, but in fact I belong here as much as the Turkana woman, for the missing links I seek aren't just *her* ancestors, they are yours, mine, everyone's. It may be her home now, but this bleak and beautiful desert preserves all that is left of the homeland of every human being on earth.

Here, I embarked on an enigmatic journey that led me to the fossil that told us more about our collective past than any other specimen ever has. This remarkable specimen directed me along a new route, one that led me into the past to find a new understanding of our ancestors and pushed me into the future of scientific inquiry, pointing to new questions about the fundamental nature of being human. He—for this curious fossil was a male, a young boy—overturned known "facts" in paleoanthropology with little regard for venerable ideas and gray-bearded wisdom. Along the way, I found many links that had once been missing—and a few that had been found years ago, but were missing all the same.

1

THE YOUTH, STONE DEAD

August 22, 1984: Mac wasn't high-grading; he never does. Searching only in the most promising areas isn't the key to his success; perseverance is. He walks the same territory over and over again, changing courses around obstacles, and he tells his people to do the same. If you walked to the left around this bush yesterday, then walk to the right today. If you walked into the sun yesterday, then walk with the sun at your back today. And most of all, walk, and walk, and walk, and *look* while you are doing it. Don't daydream; don't scan the horizon for shade; ignore the burning sun even when the temperature reaches 135°F. Keep your eyes on the ground searching for that elusive sliver of bone or gleaming tooth that is not just any old animal, fossilized and turning to rubble, but a hominid. Those are the prizes we seek; those are the messengers from the past.

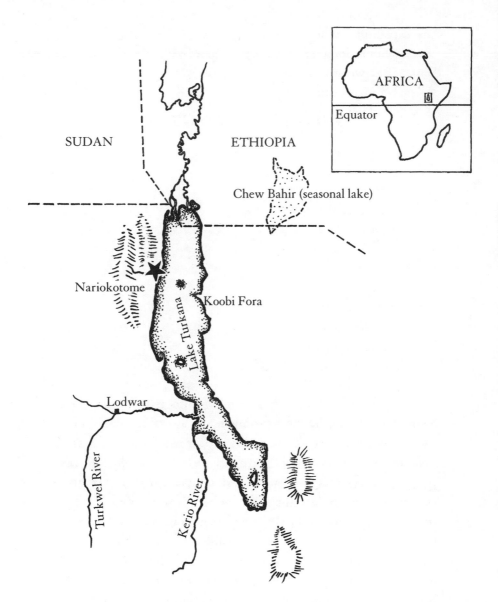

The Lake Turkana region

Mac is Kamoya Kimeu, the foreman of Richard Leakey's and my expeditions when we were searching for fossil humans together. Some long-ago graduate student gave him this nickname, and we still use it occasionally. He's one of my closest friends, a man I can work with day after day under trying conditions and never quarrel with, a man I trust my life to on a regular basis. Mac is the man who has led the team that has found more primate and human fossils than any group of professional paleoanthropologists anywhere, which is why the National Geographic Society awarded him the John Oliver LaGorce Medal in 1985. That medal has been won by such impressive people as Amelia Earhart and polar explorer Commander Robert Peary. It was presented to Kamoya by Ronald Reagan in a White House ceremony, not a bad accomplishment for a self-made man, the son of a M'kamba subsistence farmer with too many children and too little money. Mac, who has had a little schooling, has worked for the Leakeys—first Louis and Mary, at Olduvai, in the 1960s, and then their son Richard—for a long time. He is dead honest and a wonderful manager of people, well respected by both his tribesmen and his co-workers from the National Museums of Kenya and from other institutions around the world. He is a solid person, chunky and strong of body, equally immovable on matters of right and wrong.

He recruited and trained the entire hominid gang too—the half dozen gifted Kenyans he taught to find fossils. The gang are mostly Wakamba people, like Kamoya, brothers and cousins and relatives of someone who knows him in his home district of Ukambani. These are Kamoya's "people," his professional family for whom he is responsible, in whose work he takes justifiable pride. For, most importantly, Mac can find fossils where nobody else can, and he trains his people to do the same.

Some of those skills, like finding a hominid fossil, are fairly intangible. Learning to recognize a hominid simply requires training in osteology, and Kamoya makes sure the gang knows the bones of any animal they are likely to encounter. But *finding* a hominid fossil requires something else. Seeing one for the first time, I always have a feeling that I have known at the back of my mind that it was there. It is more than a search image or a mental template of its shape that is fulfilled by the reality; it is a discovery of what I already knew. The fossils always look as if they belong there, in place. The few times I have been shown a fossil that was moved before I got there, I have always known it. I can sense the place the fossil has made for itself on the outcrop by shading the ground from the sun and keeping off the rain or wind. That's what Mac must have sensed on that August day: that rightness, the belonging, the elusive feeling that you have found something in its place, something that in this case was of tremendous importance although it can't have seemed like that at the time.

The preamble to his discovery was pretty normal for our expeditions. We were excited to be starting to work a new area, the west side of Lake Turkana. Frank Brown, our bearded geologist from the University of Utah, was mapping the geology and collecting samples for dating. John Harris, our paleontologist who works at the Los Angeles County Museum, started documenting the fossil faunas, or suites of animals, that once lived together. When he compiled these into a changing evolutionary record, John could "read" the shifts in habitat and climate from the changes in the species as they evolved. For example, if several animals evolved adaptations for aridity, it would lead him to one conclusion, while the opposite would be suggested if open-country species gave way to others that preferred wetter, more forested habitat. Such a difference would be seen today if the wildebeests of the dry Serengeti Plain were replaced by an antelope that loves wet, forested areas, such as the waterbuck.

Kamoya and the gang drove up with the lorry to set up camp in a promising area. They scattered the tents among a grove of acacias, for shade, and dug a water hole about one hundred yards upstream in the sand river. We don't use the locals' water holes. An extra ten people could run the water hole dry and leave the inhabitants with nothing. The gang also built an airstrip, so Richard and I could fly up with supplies or join them when they started finding hominids. "Building an airstrip" means finding a somewhat level spot of the right length, clearing away any large bushes or small trees, knocking down any anthills, filling in the potholes, and marking the strip with a series of rocks covered in bright white plaster, so Richard could see it from the air. Bush airstrips bear little resemblance to what most people think of as an airstrip, but they can be crucial if someone is ill or injured.

The gang soon began to find many fossils in sediments that Frank said ranged from about one million to almost four million years ago—a great and interesting time period during which massive evolution occurred as hominids went from a very early, apelike species to a much more human one, *Homo erectus*. (Knowing where the fossils fit into the stratigraphic framework—the sequence of geological beds—establishes their dates, because the techniques we use at these ancient sites can be applied only to the rocks themselves, and not to the fossils.) But the fossils they kept finding weren't hominids. We always find dozens more crocodiles, hippos, bovids (antelope and their kin), and giraffes than hominids or even monkeys. Our ancestors and ancient relatives were rare creatures. Still, Kamoya was convinced that the hominids—homs, as we call them in our own slang—were there; he just had to find them. Things usually go slowly at first in a new area, so Richard and I stayed in Nairobi doing lab and office work and waiting for Kamoya to call on the radiophone. He always checks in twice weekly, for safety, and he calls sooner if there is good reason.

August 22 provided an excellent reason for an unscheduled call. The next day, if they didn't start finding homs, the gang were going to move the camp from the sand river called Nariokotome, a Turkana name that is pronounced as a six-syllable word (*na*-ri-o-*kot*-o-me). It was supposed to be a rest day—the first the gang had had in two weeks—but Kamoya was restless and dissatisfied; they had found fossils, but no homs, nothing good. The failure hurt his professional pride in himself and his team. The others were resting, doing their laundry, writing letters home: taking it easy and trying to forget the nagging calls of the fossils they hadn't found. But Kamoya started walking, "just walking," he said later, "because other people were resting. We had come far, very far; we were very tired."

Not finding good fossils for so long in a good area breeds its own sort of frustration and fatigue. "You know, to walk continuously two weeks is too much," he remembered later. "It's all you can do, you go crazy, you can't think. So while they were resting, I just walked across to see how the country looked." He went to an improbable place, a little hill on the opposite bank of the sand river, near a small acacia tree and a good-size salvadora tree. The Turkana were always taking their goats and camels through there, and little children often spent most of the day up in the salvadora, eating its pungent, sweet-sour berries instead of looking after the goats. Turkana think nothing of sending one or two boys under the age of ten off on their own to look after the herds all day, and the boys are always hungry. Not only had the place been trampled and scuffed and walked over, but it wasn't much of an erosional surface anyway, not a place to find fossils.

Kamoya went anyway. And amidst the litter of black lava pebbles and dried leaves and sticks, he found a piece of hominid frontal bone. It was the size of a matchbook and the color of the pebbles. Lord knows how he saw it. He picked it up because it was lying loose on the surface, and then he turned it over. To him, it was obviously a fragment of bone from the cranial vault, the bony covering of the brain. The inside was smooth, from the impression of a large brain—not as it would be on a pig or a gazelle, but as it would be on a hominid. He knew from the thickness that it was *Homo erectus,* the species that immediately precedes modern humans, the one that has always been identified as the missing link. It is a species we thought we knew a lot about, a foolish illusion that would be shattered by this little scrap of skull and the other bones that soon followed.

Kamoya put the fragment carefully into a small plastic box for safekeeping and returned to camp to wake up everyone with the good news. Then the entire gang came across and searched for more bits, treading carefully, disturbing nothing and finding nothing. Kamoya's piece wasn't too impressive, but it was *something.* He called us on the radiophone and we flew up the next day, on my forty-sixth birthday.

Standard procedure on our expeditions is that you never pick up a fossil hominid if it is still in situ, in its place in the rock, unless it is in obvious danger of destruction. It is far too easy to snap off a protruding piece, leaving part of the bone deep in the rock, or to forget the fossil's exact location. Anyone who finds a hominid marks the spot with a small cairn of stones and tells Kamoya, who gets Richard to excavate it, or me if he is unavailable.

When he was team leader, Richard took responsibility for the condition of the fossils, and he didn't want anything damaged unwittingly or through haste, so he usually did the trickiest work himself. Richard is the man most people credit with Kamoya's discoveries. He is the inheritor of the famous Leakey name, the celebrity whose face was on the cover of *Time* magazine, the man who can hold an audience of thousands spellbound while he tells them of the mysteries of human evolution. He started fossil finding with Kamoya in the 1960s, when Richard was an intense, very young man of the type familiar in Kenya. Such young men can do anything practical, but lack polish and academic training; they are usually more than a bit provincial in outlook and experience; they are full of confidence. They are generally great people in the field. Back then, the men nicknamed Richard *mbui,* meaning "ostrich," for his long-legged gait in walking over the exposures. Time and experience turned Richard from a gawky youth into a self-confident, cosmopolitan man, proud of his subtle French cooking and confident of his thorough knowledge of paleoanthropology despite his complete lack of university training. "I was trained," he explained once in a public lecture, "in an institution of higher learning that grants no degrees: my parents' home."

"Mac found a bit of hom today . . . ," I wrote my wife before leaving Nairobi. As I scribbled the words, I was pretty sure this fossil wouldn't lead to much, since the fragment was so small and the gang had already established that there was nothing more on the surface. If Richard and I had not been optimistic at the sound of this find, our hearts sank when we saw the small fossil, a rectangular piece about one inch by two inches, and the wretched little slope on the opposite side of the river. The men were in good spirits—their bad luck had broken, they felt—but we all knew that we had followed up a hundred scrappy finds like this and found nothing more. Nonetheless, with the fresh provisions that we had brought up, we dined in style, a white tablecloth and fresh folded napkins on the mess table looking out at the empty river. Two of Richard's inviolable rules are that the mess tent have a view and that the dining table be level.

In the cool darkness after dinner, Kamoya, Richard, and I sat by lantern light and planned the next day. It would be the beginning of a grueling and thankless task: sieving the new site. First we had to clear every twig, leaf, and rock from a large area on the slope surrounding the spot where Kamoya found the fossil. Then the whole gang would take Olduvai picks—six-inch nails embedded in curved, hand-carved wooden handles—and break up the top few inches of the surface layer of pebbles and disturbed sediment. How deep we had to go depended on how loose or consolidated the sediment was and whether or not any more pieces were recovered in the process. The idea was to try to discover if any more of the same skull was preserved and, if so, what level it was eroding out of. Once we had loosened the sediments and picked through the rubble, what remained was a pile of broken-up sediment that we had already looked through: backdirt. There would be a great heap of backdirt and it would have to be moved, which is where the schoolboys came in.

We always hire one or two local people to help on our expeditions, and this time our timing was especially fortuitous for them. The Turkana region, a semidesert in the best of times, had been suffering from an unusually prolonged drought of several years' duration. For a year or more, the local villages had been swelling daily with migrants from the north, pastoralists who now had no food or water for their scrawny herds of cattle, sheep, goats, or camels. Soon, predictably, the animals began to die, an unthinkable disaster. You simply cannot be a Turkana without herd animals; they are an essential element in the Turkana identity. Various charities, including one the locals called the "Salivation Army," tried to feed and clothe the refugees, but they could not give the Turkana back their herds, which was all they really wanted.

We couldn't restore their herds either, but we could offer jobs and training, especially to the youngsters who already spoke some English. We hired six adolescent schoolboys, paying them small wages and lots of food for running errands, carrying and fetching, and generally making themselves useful. They grew plump, and their skins soon took on a glossy sheen from all the good food; some of them started to grow. Their minds grew too, with the English practice and the new responsibilities. Kamoya and the other men also taught them concepts about daily hygiene and hard work. The boys pored over the newspapers from Nairobi, anxious to practice their reading (their school had no books) and to learn something of the outside world.

Our schoolboys soon caught on to excavation procedures. They kept eagle eyes on everyone's progress, sweeping the piles of backdirt neatly up into *kerais* (large, shallow metal bowls) and emptying *kerais* into wheelbarrows that they dumped nearby. Soon, we formed a huge mound of backdirt that the boys shoveled up and wheeled over to the sievers. Two men held a wooden-sided sieve about two feet by three feet with mesh in the bottom, into which a *kerai*-ful of backdirt was placed; they shook the sieve back and forth between them. The metal mesh is usually heavy-duty, with holes about one-quarter inch square; a second layer of mosquito netting—window screen—is placed inside the heavier, coarser mesh. Shaking the sieve is even hotter and dustier and potentially more frustrating than the rest of the procedure, especially if you think the chances of finding what you are looking for are slim. Everything that does not pass through the mesh is examined in case it is a piece of bone or tooth. The members of the gang are often more optimistic about this procedure—or perhaps they are more patient—than Richard and I are; they believe more readily that *this* is the time that something good will turn up. But even when doubts are strong, you have to sieve. You owe it to the fossils.

We knew sieving this day was going to be a deadly task so— ironically, given the way things turned out—we decided to postpone it. In the morning, we would all go off about seventeen kilometers to the south, where John Harris was working on the Kangaki River. He, too, had found a small piece of hominid frontal bone as well as some nice skulls of *Euthecodon,* an extinct long-nosed crocodile. Kamoya and the gang went to prospect a fossiliferous area that Frank Brown had discovered nearby, while Richard and I excavated a fine antelope skull and a good mandible or jaw of *Theropithecus,* a giant fossil baboon that is a particular interest of Richard's wife, Meave.

By lunchtime we were back in our own camp on the Nariokotome and procrastinating until we ran out of excuses and had to face up to the sieving. It was as tedious as we had anticipated. After two hours of unprofitable, dusty work, Richard and I exchanged looks that pleaded, *Isn't there something else we need to do?* We decided to ask Frank to show us what he had been doing. We left Wambua Mangao in charge of the sieving operation. Wambua is the strongest man of the gang, the specialist in tasks that require brute force, yet he is able to work in hardest rock with tremendous delicacy; he is responsible by disposition and thoroughly experienced. We knew he would see that the sieving was done correctly. Besides, unlike Wambua and the rest of the gang, Richard and I thought nothing more would come out of the sieves.

When we returned to camp a few hours later, the men came running, greeting us with shouts: "We've found a lot of the skull!" And they had. There was some more of the frontal bone, or forehead, the right temporal (the part of the skull that surrounds the right ear) and the right and left parietals (the vault bones that roof over the braincase). This gave us a good sense of the shape of the skull—it was clearly *Homo erectus*—and showed that, though this species is known for its thick vault bones, this individual's bones were relatively thin. We were not yet positive which bed had yielded the bones, but Frank's preliminary guess as to their age was at least 1.2 million years. When our Australian colleague Ian MacDougall dated the rock samples and Frank and his colleague Craig Feibel finished their stratigraphic studies, we determined the fossil to be 1.53 ± 0.05 million years old.

Of course, the men had sensed Richard's and my lack of faith in the likelihood of finding anything in the sieves and were laughing at having proved us wrong. We celebrated our error with a special dinner, courtesy of what might be called Leakey's Flying Catering Service. We ate smoked sailfish with capers, tomato soup, and steak with mushroom and red wine sauce. There was Beaujolais to drink, a cheese board, and coffee, followed by port. There is something incredibly luxurious about being able to have a fabulous meal like that in the middle of the desert, even if our everyday fare is less elegant. And we deserved it; we had inaugurated a new site and we had found a sizable amount of a good skull. Our spirits went from the gloom of our early conviction that the sieving would be useless to elation at this first, unexpected success. We didn't know, we didn't even dream, that we had found the first pieces of a specimen more complete—and more extraordinary—than any previously unearthed. But we were sure now that we would find more, maybe all, of the skull.

We were so sure that we called Nairobi to ask our sponsors, the National Geographic Society, to send out a photographer, David Brill. Geographic photographers are usually there when we find nothing, because they come either after a find has been made or in the hope that we will make a big discovery, which is a rare event. This time we decided to put off resuming work at Kamoya's site (now christened Nariokotome III) until Brill arrived a few days hence. By then, I had had the bright idea of water-washing the backdirt through the sieve. Rather than just shaking the sieves to break the chunks up mechanically, the water would dissolve the consolidated sediments while keeping the resultant dust down. The shower, a large canvas bag with a shower head beneath it, was hanging on a tree anyway. All we had to do was truck the material over to it and sieve there. It was a big improvement because the sievers no longer felt as if they were working in a dust storm. It was much more efficient too, even though water was a bit hard to come by.

When excavation resumed at Kamoya's site, we were immediately rewarded with more bones. While Brill snapped away like a demented grasshopper, all elbows and long legs, contorting this way and that to get the best photographic angle, pieces of the zygomatics, or cheekbones, came out, and so did the other temporal and many recognizable fragments. Richard flew off to pick up Meave and their daughters, Louise and Samira, and returned by late afternoon.

Meave, head of the department of palaeontology at the Kenya National Museums in Nairobi, is a rangy, athletic blonde, born and educated in England but now a Kenyan citizen, like Richard and their daughters. Her professional specialties are carnivores and primates. She is also my long-standing gluing partner, for we share the knack and patience required to fit fragmentary fossils back together. This task requires a detailed knowledge of anatomy—you have to recognize which bits come from which part of the body—but also a particularly acute sense of shape and color. Like me, as a child Meave had grown bored with jigsaw puzzles; like me, she used to turn the pieces upside down on the table, hiding the picture, to make it more challenging. We have sat across the table from each other for many days slowly reconstructing bones that Mother Nature has reduced to fragments.

Usually calm and matter-of-fact, Meave caught our excitement as we showed her the new skull. She and I started on the gluing while Louise and Samira, raised on camp life and fossils, alternately looked over our shoulders and checked on the excavation progress across the sand river.

There were some funny things coming out of the site, they reported: small fragments of elongated bits of bone, flat on two sides and curved in the other direction. Richard brought some over. "Walks," he said to me, gesturing to the new fragments, "what *are* these?"

"Leaks," I replied, smiling, "they're hominid ribs."

That they were ribs was obvious to us both. Richard's incredulity concerned their species. He just didn't dare to believe me. He just walked slowly back to the excavation. I knew it sounded absurd. We *never* find hominid ribs. Up until the time people started burying other people, a mere 100,000 or so years ago, ribs were among the first body parts to be crunched and munched into splinters by carnivores. Even if the carnivores don't get them—and they almost always do—ribs are not usually sturdy enough to survive the natural processes of decay and destruction to become fossilized. I knew all of this; after all, *my wife's* specialty is taphonomy, the study of the processes that destroy bones and carcasses and distort the fossil record. Be that as it may, twenty years of teaching human anatomy to medical students told me these were hominid ribs.

The next day, work at the excavation was yielding precious little. I found a piece of frontal bone and, almost simultaneously, a large yellow scorpion, annoyed at being hooked out of its underground lair. Our high spirits evaporated in the heat. It looked as if the run of bones was over. Finally Richard pointed to a level just above a small acacia tree and said, "Walker, if there is nothing more after this, we'll call it quits." I agreed that this seemed sensible. We had started at first light and had been working for almost three hours.

About 9:15, I looked up at the sound of a scuffle over where Richard, Kamoya, and Peter Nzube were working. Nzube is the Puck of the hominid gang, a small man who is always laughing and telling jokes. Even though he is a gray-haired grandfather now, with three official wives, he gets into such scrapes that sometimes the gang refuses to work with him until he mends his ways. He tends to be too quick and careless with the fossils too, a serious fault that is offset by his undeniable talent for finding tiny monkey fossils. As we watched, Richard shouted something and took a swat at Nzube, who ducked and ran off toward the sieves. Kamoya scrambled in the other direction, heading for the salvadora bush, and Richard called to me and Meave to come quickly. What he held in his hand was the left maxilla, or upper jaw, with two gleaming teeth in it; the other maxilla was stuck in the bark of the acacia tree. Nzube had excavated the maxilla, but he hadn't looked at it properly. Thinking it was just another amorphous lump of rock, he had carelessly yanked it out of the ground. Nzube, Richard, and Kamoya realized almost simultaneously what Nzube had found—and that he had risked damaging an important fossil. Fortunately, the fossil was safe, but Richard had taken it with one hand, yelling and slapping good-naturedly at Nzube with the other; Nzube had been in too much of a hurry as usual. But, also as usual, Nzube had found something terrific: now we had the face.

What's more, we realized that this level was the source of the bones. The skull had been lying upside down in the sediments, like a sort of fossil flowerpot. The seed of the tree had germinated inside the only moist spot around, the braincase. As the seedling grew, it forced the skull apart into pieces. When we cleaned the new bones, we found that the teeth were the first two molars, or cheek teeth. And we saw that there were no sockets for the last molars, the wisdom teeth—meaning that we were in possession of an adolescent *Homo erectus* skull, a rare find that was growing more complete by the minute. This was why the vault bones were a bit thin and why they had come apart so cleanly at the sutures, the points where one bone grows to meet its partner. Conveniently for paleontologists, skull sutures stay open—cartilaginous rather than bony—during childhood.

After lunch, Meave and I resumed gluing while the rest of the team continued excavating. Spirits were up again. After a while, Richard sent Nzube over—keeping him off the excavating for a bit, because he had been careless—with some more fragments of rib and a piece of zygomatic. I cleaned them, took a good look, and strolled casually over to where Richard was working. I tried to mask the surge of exultation that had sprung up when I saw the zygomatic.

"What a nice piece of zygomatic you sent over," I said, deadpan.

"Did it glue on?" Richard asked. Somehow he knew something was up.

"No," I said, "neither Meave nor I could get it to fit on the skull . . . because it's a piece of the *scapula*." I watched Richard's expression change; I didn't have to say anything more. The scapula is the shoulder blade. Like the ribs, its presence meant we had something more than a skull. We were almost afraid to say what we were thinking, out of superstition.

Up to that point, only one discovery of a truly ancient hominid had had enough body parts to be called a skeleton: that was Lucy, found in the Afar region of Ethiopia in 1974. Identified by Don Johanson, she is a specimen of *Australopithecus afarensis.* This 3-million-year-old species was believed to be the original and earliest hominid until 1994, when Tim White and colleagues found fragments of a still older species, *Ardipithecus ramidus,* in 4.4-million-year-old beds in Ethiopia. Lucy had a mandible, a handful of skull fragments, bits of ribs and a few vertebrae, and parts of most of the major bones of the arms and legs. Lucy had enough of the sacrum, and the ilium, ischium, and pubis that the entire pelvis could be reconstructed. She was a find of enormous importance, a superstar among hominid fossils, providing invaluable confirmation that our peculiar human mode of locomotion, known as bipedalism (literally, two-footedness), had evolved early in our history. Her bones also provided precious information about the size, shape, and proportions of our earliest ancestors. I knew we were all thinking the same thing, but not daring to say it out loud: Were we on the verge of finding something better than Lucy? We could almost smell success—extraordinary, unsurpassed success—but we had been left with nothing but dirt on our hands many times before when an astonishing specimen seemed to be within our grasp.

I went back to Meave and the gluing. A little later, Richard sent over a vertebra. It was eroded and broken, but it was clearly another part of the body that had been preserved. Our hopes were reaching a silent crescendo, each fossil giving more substance to a fossil hunter's fragile dream.

Very late in the day, Richard came over to where Meave and I were working. "I've come for a cigarette," he said. Meave and I both looked up. Richard doesn't smoke cigarettes, although he

used to smoke a pipe. He was referring to a rule of one of my paleontology teachers from my student days. "When you find
something difficult," my old prof would intone, meaning a tricky
or delicate fossil, "stop and have a cigarette and think about how
you're going to get it out." I went back across the sand river with
Richard to see what he had found. It was the right ischium, part
of the pelvis. When we saw that pelvis, we *knew.* We had more
than just a cranium, more even than a skull with a few ragtag bits
and pieces of other parts of the body. *It was a skeleton,* the only one
of *Homo erectus* that anyone had ever seen. Nearly one hundred
years had passed since the species was first discovered. Our scientific ancestors had spent their lives, expended their funds,
risked their health, built and sometimes derailed their careers, all
in the frustrating search for the missing link—and we had found
it. The moment of realization was sweet.

In the subsequent days, bone followed bone. Richard set the
men to clearing an even larger area for excavation, pickaxing off
the overburden (the overlying rock with no fossils in it) to reach
a level about a foot above the fossiliferous one. It sounds like un-
skilled labor, and it is, except when there may be precious and ir-
replaceable fossils underneath. It is the sort of task that the
hominid gang refuses to let me or Richard do. We aren't good
enough at it. Big Wambua and Musa Kyeva, probably the most
taciturn man on the gang, moved in silently to take off the overburden the way it should be done, skillfully and precisely. Even
though the excavating crew was reduced, there were more ribs,
more vertebrae. The next day, Richard found ribs, part of the
sacrum, and the bottom end of the tibia, the big bone of the
lower leg. No one else found a thing. "It's like going fishing with
my father," I complained, joking.

Like the bones of the cranium, the tibia was immature, of
course. The long bones of the arms and legs develop from at
least three centers: a proximal, or top, end, like the surface where
the tibia meets the femur at the knee; a shaft, or the cylindrical

middle piece; and a distal, or bottom, piece, where the tibia meets the anklebones. The small bony growth centers at the top and bottom are called epiphyses; eventually, each epiphysis will fuse with the shaft, when bony maturation is reached and growth is finished. Until then, only a plate of cartilage attaches the epiphyses to the shaft, so the bones of a dead adolescent fall to pieces easily.

By studying X rays of living people, biologists have documented the sequence in which different epiphyses fuse. It is fairly easy to determine exactly how old a youngster was at death by looking at which epiphyses are fused and which are unfused. If we kept finding bones and epiphyses, we were going to be able to pin down the age of our *Homo erectus* precisely, to within a few months. This would be useful information, since age at death might give us some insight into whether *Homo erectus* matured slowly, like humans, or rapidly, like chimps.

On the afternoon of September 1, Richard had to fly back to Nairobi briefly, though it was killing him to leave. Transparently, he suggested all sorts of tasks we really ought to attend to, jobs that would delay us from finding any more of the skeleton until he got back on the fourth. (He hates to miss a good find.)

When he returned, he decided to fly his mother, Mary, up too. Though she was then in her seventies, Mary has spent a lifetime on digs and is as tough and mentally sharp as ever. She, too, would hate to miss a good find, and we had a comfortable camp where conditions were far from rigorous—in fact, they were luxurious compared with the way she and Richard's father, Louis, had lived at Olduvai.

Since my wife, Pat, was stuck at home in Baltimore, a long way away, I asked Richard to send her a cable while he was in Nairobi. "Be advised," his historic telegram began, "that we are in the process of excavating an *erectus* skeleton." Those few words would be enough to tell Pat that this was an unprecedented find, the best thing we had ever discovered in our highly successful fifteen-year collaboration. Such an ancient skeleton would make headlines. Better still, it would provide the raw material for a hundred studies of how *Homo erectus* grew, lived, ate, and moved that would answer a thousand questions about that species' life and adaptations we had never before been able to formulate, much less resolve. Unable to sustain such emotionless formality, Richard concluded, "It's fantastic and Walks wanted you to be amongst the first to know." Pat whooped and hollered when she got the telegram, knowing the enthusiasm that a wonderful field season brings; she was also lonely at missing it.

We did have an outstanding field season. Despite carpet vipers (extremely poisonous) in the mess tent and a scorpion sting that made one of the gang miserable for two days, we had surpassed ourselves repeatedly. We had the femurs (thighbones), their epiphyses, and both tibias; incredibly, there were some *complete* ribs, curling up out of the sediments; and we excavated both intact clavicles (collarbones), the humerus (upper arm bone), and more scapular bits. We also hit upon a little cache of straight-rooted teeth that had slipped out of their sockets after death. On one of the last days, we found the mandible, lying teeth down. We could see by now that we had an adolescent boy; his browridges were fairly well developed for a youngster, meaning that, had he lived to grow up, he would have had the hulking browridges that decorate male *Homo erectus* foreheads.

And, we thought naïvely, he could talk, that most human of characteristics. This was no surprise, except to the Leakey girls. As they hung over the backs of Meave's chair and mine as we glued, I reached a crucial point and said to Samira, "Give me your finger." She reached her hand out, and I placed it into a thumb-size hollow on the inside of the boy's left temple.

"Do you know what that is?" I asked.

She shook her head no.

"That's for Broca's area, the region of the brain that controls the muscles you use in speaking."

She snatched her hand back, giggling. As she and her sister ran off to tell their father, their words drifted back to us: "The boy could *talk!*" and "What do you think he would say?"

Looking back, this incident stands out in my mind vividly. I guess at that time I wasn't listening carefully, because I didn't hear the faint crackling and creaking that signals the imminent breakup of a long-cherished idea. I should have. Language, a topic that had never been discussed much with reference to *Homo erectus*, would soon become one of the most significant issues of all.

As I had explained to David Brill, our photographer, over dinner the night before, the skulls seem to talk to those who are sympathetic. When you search for them and sweat over them and love them, you begin to hear what they are saying. The men joke that what they speak is Kikishwa, a made-up Swahili word that means "the language of the skulls." Brill asked Kamoya if skulls speak to him, if that is why he is so good at finding them.

"Yes," he said, nodding, and then fell silent a moment, thinking how to express the truth. "But," he continued with an earnest frown, "you can't understand them!" We all laughed so hard at this exquisitely accurate observation that we feared our rickety camp chairs would collapse.

For the following few weeks, the excavating brought nearly nonstop excitement, but there was some meticulous scientific work behind the celebrations. We made a detailed map of the excavation, recording the position of each bone to the nearest centimeter in three dimensions and drawing its orientation in the ground on the map. The bones kept coming, right up to the last moment, so we knew we would have to come back. Nearly everything we found was part of our skeleton. There were just a few bits of hippo, and a large, rounded footprint where that hippo or another had stepped onto the humerus and squashed it down into the scapula, pushing both into the mud that preserved them; and the opercula, or shell-like trapdoors, of a species of freshwater snail called *Pila*. Since the fossils were scattered over a fairly wide area and we were not yet sure what had happened so long ago when this individual died, we were going to have to make a *huge* excavation when we returned, in order to be certain of finding all of the skeleton's bones that were in the ground.

When we closed down the site for the season, on September 21, 1984, we had found more of *Homo erectus*—the classic missing link—than anyone had ever seen. The next four field seasons laboring in the pit, as we came to call the enormous excavation, would see 1,500 cubic yards of rock and earth moved by hand. Our schoolboys, who worked with us faithfully year after year, grew from adolescents to young men while the Nariokotome boy, as we took to calling the specimen, grew from a fragment of skull to the most complete early hominid skeleton ever found. By 1988, for the first time in history, we were able to look at an almost complete skeleton—not just a scrap of skull, a handful of teeth, or a portion of an arm, but a bony record of one individual's life. It was an extraordinary opportunity.

We stopped excavating because, during the final two seasons of digging, we found only two scraps of the hom and more hippo, tortoise, and catfish bones than we wanted. Our lack of success wasn't due to lack of help; every paleoanthropologist passing through Kenya wanted to spend a few days digging such an important site. Inch after inch, *kerai* after *kerai,* wheelbarrow after wheelbarrow, neither the sievers nor the excavators found anything new. Eventually, it simply seemed as if we had found everything that was left of the boy. Looking at the site plan, my wife, Pat, thought the still-missing bones of the hands and feet probably were no longer in the hillside. Her guess was that they had eroded out first and been pounded into dust by Turkana goats' pointy little feet. Mary Leakey disagreed, saying we should keep digging until we found all of the Nariokotome boy. After all, she and Louis had worked thirty years at Olduvai before they found their first good hominid. Richard and I considered the tens of thousands of dollars that each year of excavation was costing and the seemingly endless man-hours of work that we had already expended, and decided to quit.

Still, by autumn of 1988, our boy was hauntingly complete, an adolescent who we guessed was about twelve or thirteen years old, roughly the age of the young Turkana kids when they had started working in our camp. Like one of them, the boy was about to lose the milk canine or eyetooth from his upper jaw. It was at that wobbly and irritating stage where you keep fiddling at it with your tongue but it hurts too much to pull out. Next to it in his jaw we could see the permanent canine, well formed and ready to erupt. But he had died first, lying face down in this swampy little pond, his carcass kicked and stomped by hippos, his bones nibbled by catfish and buried in inches of slimy mud. Except for parts of the feet and hands, and some of the vertebrae and fragile ribs, everything else was there, preserved in rock, saved for us.

There was no obvious cause of death—there rarely is with fossils. The only hints were the signs of periodontal disease that had damaged the bone on the right side of the lower jaw, next to one of the permanent premolars, the tooth that lies two to the rear of the canine. When we took radiographs of the jaw, we could see what had happened. A few weeks or perhaps a month before the boy died, this permanent premolar had replaced the second milk molar. (Because the deciduous teeth are working when the jaw is small, they are fewer than adult teeth; permanent premolars replace milk molars, and later the adult molars come in to their rear.) But even though the premolar was ready and moving upward to erupt through the gum, the roots of the milk molar hadn't finished resorbing yet, an occurrence that is still fairly common today. When the milk molar fell out, two fragments of root remained behind in the jaw, with channels through the gum that now opened to the outside. Not surprisingly, infection set in and an abscess formed. Maybe it was enough to kill our boy: we will never know for sure. It sounds trivial but, even as recently as Shakespeare's time, "teeth" were listed as a leading cause of mortality, and dental infections were deadly serious.

We knew by then that he *was* a boy, but he still held a thousand secrets locked in his gleaming teeth and gently curving bones. He had seen a world we could glimpse only if he would speak to us. *He* had captured our futures. There were masses of work yet to do in the laboratory as we entered this uncharted territory. Who could have guessed the boy would overturn long-held beliefs, bulwarks of paleoanthropology's implicit paradigms?

We all knew that *Homo erectus* was a pretty human sort of creature from the neck down—a bit short, stocky, and muscular, the type that makes a good soccer or lacrosse player. It was only the funny, big-browed head containing a small (by human standards) brain that was different. In Richard's television series, *The Making of Mankind,* actors representing *erectus* had worn elaborate masks on their heads, but needed no body suits to be convincing. Of course, *erectus* had stone tools—hominids have had stone tools since about 2.4 million years ago, well before *erectus*—and we knew the species had language, which probably accounted for its survival and evolution into big-brained humans.

Didn't we?

2

A SKELETON
IN THE CLOSET

The irony of paleoanthropology is that most of the links are missing. By 1988, the skeleton of the Nariokotome boy was so complete that he became a showstopper whose photo earned me or Richard a round of applause at every lecture we gave. He *looked* so human—you didn't need to be an anthropologist or an anatomist to see he belonged to a species ancestral to us. People were thrilled to realize he is 1.5 million years old. Funnily enough, I myself have never been much taken with this missing link: it is entirely too human for my tastes. At one point, I remember thinking what a shame it wasn't the skeleton of a more interesting species. But of course the boy proved more interesting than anything else we have found. Before him, not enough was known to reconstruct a *Homo erectus* skeleton accurately, even though paleontologists and anatomists had been seeking his kind since the nineteenth century. Previous discoveries of his species had included only skulls, jaws, and scraps of arm or leg bones, whereas the Nariokotome boy was a wonder of completeness.

All our specimens are assigned catalogue numbers, which we often use as nicknames. In honor of the boy's extraordinary completeness, we looked for an upcoming number that was euphonious. He became KNM-WT 15000. KNM stood for Kenya National Museums; WT for West Turkana, the region where he was found. But 15000 — 15K as we abbreviated it—was his and his alone.

I should have foreseen that surprises were in store, especially since strange twists and turns of fate, and of ideas, are endemic to the story of the missing link. In fact, the very notion of a missing link is mesmerizing—why should anyone think there should be such a thing? It didn't really come from Darwin. When he published *On the Origin of Species* in 1859, he had no direct knowledge of human evolution. (The only known fossil hominid, a Neandertal, had been discovered in 1856, but reports of it had not been translated into English.) Although Darwin knew perfectly well that evolutionary theory applied to humans, he deliberately sidestepped this most controversial issue. His theory of evolution was disruptive enough without it. "It is like confessing a murder," he wrote of the anguish of explaining his beliefs. His champion, Thomas Henry Huxley, however, attacked the problem head-on in the famous debate against Bishop Samuel Wilberforce and in his 1863 book, *Evidence as to Man's Place in Nature*. Darwin shied away from dealing directly with the issue until 1871, when he published *The Descent of Man and Selection in Relation to Sex*.

Yet the term *missing link* seems to have been on everyone's mind and lips even if Darwin avoided it. It was generally taken to mean a transitional form that was half-ape, half-human, the link that would prove evolution. The concept existed, vaguely, even before *Origin of Species* was published. For example, in 1855, Richard Owen, an archenemy of Huxley, and later of Darwinian evolution too, had exaggerated and misinterpreted the structural differences between apes and humans, referring to the "last link in the chain of changes" between them, a link he couldn't find. This was a play on the Great Chain of Being, a medieval concept, with origins dating back to Aristotle. According to this notion, all creatures were arrayed as a chain of creation that proceeded, step by gradual step, toward perfection. This metaphorical chain linked the lowest beast to the highest, mankind, which was in turn linked to angels, archangels, and the Divine Being Himself. The chain did not represent evolution, Darwin's "descent with modification." It represented instead a static differentiation, a sequence or orderliness to the objects of creation, not change.

But the concept of a missing link became associated with Darwinism, however Darwin felt about it. In 1877, the students of Cambridge made a visual pun on it as they decorated the Senate House for the ceremony at which Darwin was to be awarded an honorary degree. During the afternoon, Darwin's wife, Emma, kept glancing toward some cords strung from one gallery to the other, waiting for something to happen. She was not disappointed. First a monkey marionette appeared, causing great uproar until it was removed by a proctor, and then "a sort of ring tied with ribbons which we conjectured to be the 'Missing Link.'"

The missing link was refined from a general concept to *Homo erectus* in particular because of the work of two extraordinary men. The first was Ernst Haeckel, the prominent German biologist and a great admirer of Darwin, who promoted evolutionary theory shamelessly in Germany in the face of fierce opposition from his former professor Rudolf Virchow, who was probably the most powerful biologist in Germany in the late nineteenth century. Virchow was the ultimate data man, eye to the microscope, obsessed with facts and not theories. Haeckel was the opposite, a man of broad vision who loved to soar beyond the facts into the heady realm of theoretical speculation. He was, in fact, a trifle careless with the facts from time to time. In 1868, in a move that probably gave Virchow apoplexy, Haeckel proposed (on purely theoretical grounds) a new species of fossil human called the missing link. This link, half-ape, half-man, was boldly named *Pithecanthropus alalus* and was drawn in on Haeckel's diagrams of the human evolutionary tree, a handy metaphor that Haeckel also introduced. The name meant ape- (*pithec-*) man (*anthropus*) without speech (*alalus*). Speech was the crucial attribute Haeckel expected this ancestor would be lacking, though he hadn't a shred of proof; this aspect of his prediction was largely ignored. It wasn't until we had analyzed the skeleton of 15K more than 120 years later that the true significance of Haeckel's intuition became apparent.

Haeckel also believed, wrongly, that the Asian apes (orangutans and, especially, gibbons) were more closely related to humans than were African apes, so he confidently drew a line indicating the descent of this missing link from the branch that housed the Asian apes. He expected the missing link to be primitive, so he described it as hairy, with a long skull and slanting or protruding teeth; he also predicted such creatures would walk half-erect.

If Haeckel's phylogeny—another term he coined—was nearly fossil-free (Neandertals were the only nonmodern hominid then known), neither the lack of evidence nor Virchow's vigorous opposition stopped his ideas from becoming enormously influential. Haeckel had a gift for speaking and writing and a romantic penchant for generating grand, sweeping theories that made him an immense popular success. His books, accessible and persuasive, were translated into many languages, and hundreds of thousands of copies were sold to an eager public—despite the fact that, in many ways, the books were wrong.

One of those who devoured Haeckel's work was an intense young man from Holland, Eugène Dubois. It was he who brought the missing link to life, so to speak. When I met up with the missing link myself, at Nariokotome, I could not help but think of Dubois, whose life was shaped and ultimately haunted by finding the missing link. Mine is a historical science, and I couldn't study *my* missing link without remembering the others and their discoverers, starting with Dubois. I have come to admire and sympathize with him, and to see him as a man of courage. While a young medical student, he read Haeckel's words and took as his lifelong ambition the capturing of the missing link. From early childhood, Dubois, like me, had been fascinated by fossils, scrabbling through quarries and road cuts to find them. Once he knew of *Pithecanthropus*, he became obsessed with finding it and winning scientific fame and fortune. He heard the past calling him like a Lorelei, pulling him out of his ordinary life into dangerous waters where something extraordinary might happen.

Here we differed. My aim never was to find the missing link but rather to study fossil fish. When I finished at Cambridge in 1962, I was awarded a grant for graduate study. But I needed a supervisor for my planned Ph.D. on the evolution of fish. A friend introduced me to Kenneth Oakley, an anthropologist at the British Museum (Natural History), now known as the Natural History Museum. The BM, as it was affectionately known, houses a wonderful collection of fossil fish and at that time employed Errol White, a logical man to be my supervisor. Kenneth introduced me to White, with whom I made an appointment to discuss my future.

At exactly the appointed hour, I presented myself at his door. (I am always on time.) Because White's office faced the public galleries, the door to his laboratory and office suite was always locked. I knocked politely, and waited.

"Come in!" a voice called gruffly.

I tried the door. It was locked. I waited. Then I knocked again.

The exchange repeated itself fruitlessly.

Eventually, White stopped answering and I stopped knocking. I stood there helplessly, in mounting frustration, for half an hour until another curator happened by and let me in with her keys. I walked into White's office.

"Who are *you?*" he growled, looking up.

"I'm Walker," I said.

"You're late!" he replied, angrily. "It's rude not to be on time."

"Well, it's bloody rude of you to keep me standing in the gallery, when you know I don't have keys," I answered, furious but not raising my voice. It was not a propitious start to a future relationship.

He glared. "Now that you're here," he said, "why don't you tell me what you want?"

I took a deep breath and started. I had devised three different thesis projects, all interesting and all feasible, I thought. To the first one he said, "No. Someone else is already doing that." To the second one he said, "No. I'm saving that for my old age." And to the third he said, simply, "No."

I looked at him for a moment and then left, despairing. Our conflict was not only between a young and ambitious man, anxious to make his place in the world, and an older one, guarding his resources and position jealously. It was also the classic English confrontation, where someone of the working class failed to address his elder and social superior with what the latter regarded as the proper deference and humility. I had seen it many, many times before, and repetition did not blunt my dislike of the situation. I simply wasn't going to tug my forelock for anyone. I had been smart enough to come top of nearly every exam at my grammar school and win a scholarship to Cambridge; I had proven my abilities there over and over, earning a good degree and a grant for graduate study. I couldn't see that it mattered what my father did for a living or who my mother's ancestors were.

But, of course, by getting right up White's nose (not to mention that he got up mine too), I had landed myself with a serious problem. I went back to Kenneth's office and told him what a debacle my interview had been. If I couldn't work with White, I couldn't work on fossil fishes. What was I going to do my graduate research *on?* Did Kenneth have any fossils?

"No," he started to say. "Nothing that would make a thesis. . . ." Then he remembered that there were some fossil lemurs in some drawers in the basement. He thought we should go look. *Lemurs?* I asked myself silently as we wound through the corridors and stairways, *What can I remember about lemurs?* All I could conjure up was an image of the ring-tailed lemur, the fox-faced primate from Madagascar that I had seen in the London Zoo. As we opened drawer after dusty drawer, there seemed to be plenty of bones: beautiful skulls, limbs, and pelvises of varying shapes and sizes. They were what are called subfossils: extinct species, but not extinct for so long that their bones had yet been completely fossilized. A quick check established that no one had done any substantial work on them in thirty years; it was time for a new look. But Kenneth wasn't the right person to supervise me, as his expertise was in chemical techniques for dating fossil man; he suggested instead John Napier, an anatomist at the Royal Free Hospital Medical School.

It was a lucky choice for me, for no one ever had a more exciting time as a graduate student than I did. John made every day an adventure; he was a brilliant teacher who was at the hub of primate evolutionary anatomy. We were discovering the rules that linked locomotion—movement—with anatomy and that related both to primate evolution. John was collaborating with Louis Leakey, trying to apply what had been discovered about locomotion in living primates to the fossil record that Louis and Mary were digging up at Olduvai Gorge in Tanzania. It was an exhilarating time when primatology was just starting to develop, both as an important specialty within animal behavior and as a tool that illuminated the fossil record. Everything and everyone of importance seemed to pass through John's lab, and every new idea was eagerly discussed. For three years, I woke up each morning longing to get to work and learn something new.

Dubois's experience was very different, though his youth and impatience seem familiar to me. I smiled ruefully when I read what one of his colleagues wrote: "Dubois had the habit of just lifting a corner of the veil of a scientific concept, but he was loath to settle down and work it out thoroughly; he preferred to leave to others with more perseverance the task of continuing and finishing it." On occasion, my colleagues and students accuse me fairly of getting bored once I have 90 percent of the answer to a question.

After medical school, Dubois was drawn to research in comparative anatomy and, like me, worked with a leader in the field, in his case Max Fürbringer. Dubois took the job as Fürbringer's subordinate and heir apparent, but the role was not easy for Dubois, a prickly, proud individual, easily offended. He even asked Fürbringer not to attend his first lecture at the University of Amsterdam for fear his questions or comments afterward might make it appear as if Dubois were not yet standing on his own. Fürbringer was a bit nonplussed, and wrote to Dubois: "So the question arises whether I am at least permitted to attend the lecture with a muzzle, or whether you are afraid that even this might damage your independent status." The tensions between them heightened until, in 1887, Dubois at the age of twenty-nine suddenly quit the university and sailed for the Dutch East Indies with his young wife, Anna, and their baby. At that time, he was harboring a grudge against his former adviser over the authorship of a paper based on work they had done together. He refused to tell Fürbringer when their ship was sailing, even though Fürbringer had asked to see

them off. Dubois simply left a letter that was delivered to Fürbringer after the Dubois family had departed. It was the first of
many fallings-out with friends and mentors during Dubois's
career. Sometimes the rifts healed quickly; other times they
were fatal to the relationship.

In this case, Dubois undoubtedly chafed at the slowness of his
advancement in academia; indeed, the transition from student
to colleague is often an unhappy one. He was always fiercely
protective of his intellectual rights and often feared people were
stealing his ideas. But another man in his circumstances would
have found no cause to leave Amsterdam, much less to go as far
as Indonesia. Dubois was not simply escaping his situation: he
was risking all for the chance of finding the missing link.

He signed on with the Royal Dutch East Indies Army as a
medical officer. He was convinced from his reading that the fossils that told of human evolution would be found in the tropics.
Wasn't that where all the living apes were? And weren't the
tropics spared the grinding destruction of the glaciers that had
scoured Europe during the ice ages? If Neandertals had inhabited Europe, as he knew from recent finds they had, then the
earliest human ancestors, the true missing links, had lived farther south. The East Indies, as a Dutch colony, was one of the
few places where it was feasible for him to pursue his quarry.

Still, it was a leap into the unknown. At a similar stage, when
I had almost completed my Ph.D., I took a post at Makerere
University College in Uganda, with John's blessing. It was a
good job and much nearer the living and fossil primates that fascinated me. Uganda was newly independent then; the university
was terrific, full of enthusiastic faculty and students mining the
wealth of research opportunities. There was a heady feeling of
youth and optimism in the air. My first wife, Rikki, my infant
son, Simon, and I packed up and sailed on the S.S. *Rhodesia Cas-
tle* to Mombasa, on the Kenya coast, to take the train cross-
country to Kampala.

What I noticed first upon arrival in Mombasa was the blinding sunlight, reflected off the old Arabic-style buildings, making the bright colors of bougainvillea and jacaranda glow. The second thing to strike me was the air: palpable, thick, warm, and humid. It had been hot, by British standards, since Gibraltar. But here the air smelled of frangipani, jasmine, and tangy spices. People of every color and shape walked by dressed in an amazing range of garments. There were shapeless black *bui-buis,* covering Muslim women from head to toe, and brilliant, gold-bordered saris; there were Swahili men in shirts and *kikois* (striped cloths wrapped skirtlike around their hips), and the occasional, red-ochered Masai warrior, strutting in a red-checked *shuku,* or cloak, that looked as if it had been borrowed from an Italian bistro; there were Africans, Europeans, and Asians in a mixture of fashionable and old-fashioned European garments. It was nothing like gray, dismal London. It was an excitingly young and incoherent society. Africa has a history, to be sure, and strong traditions, but Kenya in those days was such a hodgepodge of diverse people and languages and cultures that anyone's particular personal history was irrelevant. Africans knew in an instant, of course, what tribe another African came from and indulged in amazing prejudices rooted in differences in custom. "That boy," a Kikuyu student might say to me after a lecture on race, pointing to a Luo student, "eats *fish.* He is very black. I am surely closer to you than to that boy." But to them, an Italian, an Englishman, or a Lebanese were practically indistinguishable—all "Europeans" without differentiation. What mattered was how you behaved and what you accomplished.

We were met, as we had been told we would be, but everything was completely confused. The man who met us had had no instructions from Kampala; contrary to my contract, there were no hotel reservations, no train tickets, no nothing. All our worldly goods were packed up in crates and suitcases, our child was small, cranky, and hot, and we had no hotel and no East

African currency. It took days to get it all straightened out, days of trudging from one office to another, of talking to endless bored officials, of cabling Kampala and waiting in the heat for an answer. Finally we got some money and took a train to the interior. The Mombasa-Nairobi train was once one of the most elegant lines in the world, and it was still extraordinary in the mid-1960s. It left Mombasa at teatime, chugging slowly away from the coast and through the darkened plains of Tsavo National Park. After a few hours, a proud African conductor in a starched white linen uniform strode through the train, striking a tune on a small xylophone, announcing the first sitting for dinner. The dining car was one of the old colonial ones, still elegant with wood paneling and brass lamps. The silverware, marked with *EAR* for East African Railways, was meticulously set for a four-course meal on a crisply ironed tablecloth. While we ate, a steward prepared our compartment, folding down the beds and making up the berths with sheets and blankets. The next morning, very early, we got our first look at the African interior somewhere around the Athi River, southeast of Nairobi; there were the golden plains interrupted by rivers that had carved their way into the volcanic soil. We had glimpses of strange and exotic birds like malachite kingfishers or carmine bee eaters, but no mammals yet. The train stopped in Nairobi at midmorning and set off again by lunch, climbing up the other side of the Rift Valley during the night. The next morning, we arrived in Kampala and settled into our university quarters, a rather ugly house with institutional furniture and a small garden. It was strange and deeply unfamiliar, but it was all a grand adventure too. By the next morning, our front lawn was covered with Africans, waiting for us to hire an *ayah*, or nanny, and perhaps a cook and a gardener too. We didn't know what to do, as we had no intention of behaving like colonial imperialists. After three days, someone explained to us that it would be very selfish, when we were so rich by African standards, not to share our wealth. It was our duty and our responsibility to hire someone. I hadn't understood this perspective before, which stood my principles on their head. So I walked out and said to the dozens of people on the lawn, "If you speak English, please stand up," and so we hired the two who did.

I have no record of Dubois's thoughts as his boat pulled into Padang, Sumatra, in December 1887, but I know this: the sheer foreignness of it all was probably as stunning to him as Mombasa was to me. Here was no tidy little civilized Amsterdam; this was a wild and sprawling place, with colonial architecture, brown-skinned people speaking strange languages, and vegetation, birds, and animals running riot in the tropical sun. At first, Dubois was stationed at a military hospital and his colleagues were medical men or military officers. He spent his own time and money looking for fossils. Soon, he badgered the East Indies Committee for Scientific Research into giving him some funding, and after two years he was freed of his medical duties altogether, to search for fossils. He was given the services of two civil engineers, and fifty convicts to excavate—a real triumph of persuasion.

The missing link would hardly have remained missing for so long if it were easy to find. Dubois soon discovered the endless difficulties of excavating for fossils in undeveloped nations. In October 1889 he wrote to the director of the National Museum of Natural History in Leiden, describing his early field seasons. There was more than a hint of the theater of the absurd, laced with tragedy.

> Everything here has gone against me, and even with the utmost effort on my part, I have not achieved a hundredth part of what I had visualized. Where cave explorations are concerned, the reverses began right at the start, with my coming here in the *poeaza* (the period of fasting) when the Malays are as indolent as frogs in winter. . . . A survey of the caves I was

provided with seemed fitted only to put me off the scent [be-cause] there were very few real caves among them. Yet these did exist, as I saw later, and people had simply concealed their existence from me . . . because they thought that the "Com-pany" would appropriate the gold and saltpetre the islanders get out of these caves. . . .

The parallel suspicion when I am excavating in Kenya is that our team is mining either gold or a mythical substance known as red mercury, described as incredibly rare and incredibly valuable. Who would believe that rich *wazungu* (white folks) would come to dig for old bones or stones of no apparent value to anyone? It makes for some interesting difficulties. This is why we often hire on local people, wherever we work. They be-come our intermediaries, explaining to their families and neighbors what we are doing, testifying from their own experi-ence that we are not secretly mining valuables. It has proven a more effective deterrent to trouble than arguing that red mer-cury doesn't exist.

Like us, Dubois also faced real dangers in searching for fos-sils, which he sometimes jokingly called "diluvialia," after the theory that these were simply leftovers from the Great Biblical Flood, rather than evidence of evolution. His letter continued:

What's more, it was necessary to live out in the forest for weeks on end, usually under an overhanging rock or in an improvised hut, and it turns out I can't stand up to that, how-ever well I was able to bear the fatigue [at first]. Having now come back, with my third bout of high fever, which nearly put paid to all the searching for "diluvialia," I have had to give it up for good. . . . The trouble with the personnel was even worse. To begin with, one of the two engineers assigned to me to supervise the forced laborers was totally useless, and after repeated warnings and exhortations to carry out his duties properly he was transferred at my request. . . . The other engineer died of fever.

Of his fifty laborers, seven ran away and half were down with fever themselves. He faced bad roads, *no* roads, overgrown forest, mountains, and rains that made travel impossible but left no potable water. He had no trained hominid gang, as we did, no Land Rovers, no sturdy tents, no modern medicines, neither airplanes nor radiophones. Even with our advantages, we have had our share of serious fevers, heatstroke and dehydration, broken bones, snake bites and scorpion stings. Sometimes there is nothing you can do except hold someone's hand through the night and wait for morning when the Flying Doctor can land. Dubois's experience was even worse, because medical care was far away and offered few effective treatments for tropical diseases. He always felt that Java had nearly killed him, writing:

> A few cool windy days did seduce me to set up my tent nearby the excavations. After a stay of fourteen days, I discovered that, because of the hellish heat and malaria, there is no more unsuitable place available in Java for this study than this hell (which I, a born Roman Catholic, merely have kept for a purgatory).

In August 1891, the sheer power of Dubois's logic and determination triumphed over the odds. He had shifted his operations to Java and had decided to stop concentrating on caves and to look at open sites. When fossils were spotted eroding out of the banks of the Solo River near the village of Trinil, he set his crew to excavating while he reconnoitered for new sites and then returned home; they were successful. They found a hominid molar, which he at first took to be an extinct ape's, a species known as *Anthropopithecus* that had been found in the Siwalik Hills in British India. Then, in October, the new engineers sent to replace the first pair found a hominid skullcap the like of which had never been seen. It was a long and low braincase, with no forehead and strong browridges like those of big male apes. The back of the skull was sharply angled, rather than being smoothly rounded as in humans. Even without measuring, Dubois could see that the brain that once filled the skull was big for an ape but small for a human. His reports show that he realized immediately that he had found the missing link.

The onset of the rainy season flooded the site. Finally, in May 1892, Dubois's crew returned to work, and in August the engineers and laborers excavated a left femur, or thighbone. The two civil engineers, corporals Kriele and de Winter, wrote immediately to Dubois, sending the fossil and describing the location of its excavation in some detail. Reluctantly, Kriele also told the ever-temperamental Dubois that the small pieces missing from the distal, or knee, end of the femur had been placed on a djati leaf and had blown away before they could be glued in place. Dubois was delighted enough with the find, it seems, not to become angry. The bone was stunningly human in size and shape, save for an obvious pathology near the upper end where some serious injury had healed, leaving a large growth of extra bone. For Dubois, there was never a moment's doubt that the three pieces (tooth, skullcap, and femur) had belonged to a single being.

The question was, what sort of creature did it represent? He continued to call it *Anthropopithecus* in his quarterly reports to his supervisor, combined with the trivial name *erectus* because the femur was so clearly from a bipedal species. This usage continued until December 1892, while Dubois waited for a chimpanzee skull to arrive from Max Weber, an anatomist in Holland, so that he could compare it carefully with the one he had found. The chimp skull came on December 18, and the next day Dubois wrote a letter of thanks to Weber in which he still referred to the specimen as *Anthropopithecus*. Sometime in the next nine days, during which time Dubois compared his skullcap to the chimp's and calculated the fossil's cranial capacity, he changed his mind about the appropriate name for his find. Amazingly, almost the exact moment of change is recorded. The Dubois Archives in Leiden include a letter to his supervisor, the director of Education, Religion, and Industry, dated December 28, 1892. In it Dubois wrote of his new species, forming first the capital letter *A,* for *Anthropopithecus,* and then overwriting it with the *P* of *Pithecanthropus.*

By now, Dubois was convinced that the only appropriate name for his new being was *Pithecanthropus* (ape-man). He decided against the second part of Haeckel's name, *alalus,* because he had neither face nor mandible to tell him if *Pithecanthropus* spoke, and the fragile braincase itself was still filled with stony matrix, so he could not search for the impression of the speech center of the brain, Broca's area. But he did have a wonderfully human femur, the shape of which contradicted Haeckel's prediction of half-erect walking. Dubois proudly gave his fossil the name *Pithecanthropus erectus.* On February 3, 1893, a local newspaper, the *Bataviaasch Nieuwsblad,* carried a semisatirical account of his discovery, taken from his November 1892 quarterly report that used the name *Anthropopithecus erectus.* Dubois clipped the article and saved it for the rest of his life. Ironically, this article is signed *Homo erectus;* it is by many years the earliest known publication that uses this name in conjunction with this fossil. But science moves slowly. In the scientific literature the suggestion to revise *Pithecanthropus erectus* to *Homo erectus* was made in the mid-1940s, but it was not until 1960 that the change was formally accepted.

During 1893, Dubois worked hard to discover exactly what his fossil meant, preparing a monograph on it in scientific isolation. He did spend long evenings talking it over with his close friend Adam Prentice, the manager of a nearby coffee plantation, who was interested and intelligent if untrained in science.

In spite of his elation at his great discovery, the end of 1892 and the first half of 1893 was a difficult, depressing time for Dubois: his father died back home in Holland, his wife miscarried, he was plagued by ill health that compounded his anxieties about his future, and he was struggling with his monograph. By July, however, Dubois's mood had lifted. Although that year's excavations at Trinil yielded no new remains of *Pithecanthropus,* he recommended Kriele and de Winter for promotion to sergeant, and photographed them standing proudly, displaying their new stripes. Dubois had resolved in his own mind that he had indeed found a missing link. By the end of the year, he had finished his monograph and triumphantly mailed it off to scholars in Europe. Early in 1894, Dubois set sail with a light heart for British India, to see *Anthropopithecus* for himself, awaiting the praise and scientific acclaim that he was certain was his due. His diaries for the first two months of his trip are upbeat and optimistic; people are charming and helpful, scenery is beautiful, natives are colorful and hardworking.

Then in early March he began receiving word of the skeptical reception of his ideas in Europe. Dubois and his monograph were openly mocked by his colleagues, few of whom shared his conviction that he had found the missing link. Scholars challenged the association of the remains, disbelieving that they had come from one individual. The standards of documentation of excavations were not as exacting in the late nineteenth century as they are now, and Dubois's engineers and convicts neglected to take any measurements of the position of finds in the ground in those early seasons, although they did write him a description of the layer in which the material was found and of its position. But nothing was mapped, and Dubois himself was not at the site when the finds were made. He also omitted geological information from his early publications about *Pithecanthropus,* failing to appreciate how important such information might prove. Dubois simply asserted that it would be foolish to believe the three fossils were *not* from one individual; his critics replied, cruelly, that it would be foolish, indeed, to think that they were.

In Germany, the battle over the theory of evolution was still bitter. Rudolf Virchow lost no time in declaring that Dubois's marvelous fossil was the skullcap of a very interesting giant gibbon, its brain (which could not be measured because the skullcap was full of rock) too small to be a human's, and the femur was quite possibly that of a human, since no ape could have sustained such a serious injury and lived while it healed. (Some anthropologists would now agree with Virchow that the femur is anatomically modern.) And yet, Virchow conceded, perhaps this was simply the femur of some extinct giant gibbon, since gibbons walk erect on the ground. In any case, it was no missing link. Virchow's numerous followers rejected Dubois's conclusions out of hand, and all of Ernst Haeckel's attempts to defend this incarnation of his hypothetical missing link were in vain.

The French were similarly scathing, with the exception of Paris anthropologist Léonce-Pierre Manouvrier. The British were condescending. If only, one English anatomist sniffed, Dubois had had an adequate collection of modern human skulls of what were then called primitive races, he would have seen that a small braincase did not exclude one from membership in the human species. The femur was also probably modern, and the tooth had nothing to do with either. Even the Dutch offered little support.

Part of the problem may have been that the Dutch government had pressed Dubois to publish his remarkable find rapidly. And he was eager to reap the acclaim he deserved; he had waited his whole life, he had risked so much, for this discovery. So he had plunged into a detailed anatomical description of the fossils even though he had few modern ape or human skulls (and no cast of the Neandertal remains from Germany) to which they could be compared. He was acutely aware of the problem and had done his best to remedy it, sending to

Holland for ape specimens and reading and taking extensive notes on every relevant publication he could get his hands on. He was a well-educated and meticulous scholar, albeit one confined to the Far East, and his excavations were more carefully conducted than he was given credit for. In truth, the problem lay more in the prevailing beliefs among his colleagues than in Dubois's shortcomings.

Accounts of his colleagues' reception, read while he was in British India far from his family and friends, changed Dubois's mood from one of confident elation to unease, misery, and bitter suspicion.

He returned to Europe in 1895 to fight back against the skeptics. He persuaded the Dutch government to give him museum space and a salary as the curator of the important and massive collection of fossils (three hundred crates' worth) from Indonesia and India. Then he went on the lecture circuit, waving diagrams of geological sections under his critics' noses and displaying the bones themselves, which have an impressive air of antiquity about them that had been poorly conveyed by the photographs in his monograph. He talked for the first time in some detail about the fauna of extinct animals found at the same site, another line of evidence for the antiquity of the *Pithecanthropus* remains. In Leiden, Liège, Brussels, Paris (on two visits), London, Edinburgh, Dublin, Berlin, and Jena, he spoke with determination bordering on pigheadedness and with conviction approaching fanaticism. He changed a few minds and even earned the friendship and respect of some of the skeptics—he was a charming man, notwithstanding his paranoia and hotheadedness—but he failed to turn the tide completely. If finding the missing link had been difficult, it was nothing compared to persuading the world that he had found it.

Even then, paleoanthropology was a contentious discipline. There is something about human evolution—maybe it's the fact that we're all human—that makes all of us feel we know something about the subject; everyone is an expert. One of my experiences shows the problem.

After an article about me and my work had appeared in a local newspaper, I got a phone call from an extremely disputatious man, who called to tell me that I simply couldn't be right that whites and blacks shared a common descent from the same ancestor. I was startled, to say the least, since there is an irrefutable mass of evidence that all modern humans comprise a single species that, of necessity, must have had a common origin. I asked him what his training was. Had he ever studied human anatomy or seen any of the fossil evidence? No, he said, he was a small-town newspaper editor. But he *knew* I was wrong. I persisted for a moment, annoyed and yet fascinated by this phenomenon. Would he walk into an operating theater and tell a neurosurgeon what to do? I asked. Of course not, the editor replied indignantly; neurosurgeons had years of training in a highly specialized field. He couldn't pretend to know anything about neurosurgery. Then why, I said, did he think he could tell me how to do my job? It had taken similar years of training and practice, and he hadn't even bothered to acquaint himself with the popular literature on the subject, much less the technical literature. He spluttered that he had a perfect right to his opinion and that it was as good as mine. It was, I said, but not about matters of fact that could be verified, not if he wouldn't base his conclusions on the evidence.

Ideas about where we came from and how we evolved cut so close to the bone because they are so intimately linked to our vision of ourselves. It can be disorienting and even threatening to be presented with a view of yourself that deviates strongly from the one you have always held. An extreme example is the rare medical condition known as the testicular feminization syndrome. Raised as females, sufferers are often identified when they seek medical help as adults because they are unable to conceive. Despite their female gender identity and female-appearing external genitalia, these patients are genetically male but have become effectively female because of their insensitivity to androgens (particular male hormones). The discovery is often profoundly disturbing to these individuals. Similarly, for someone who has spent decades thinking about the nature of our collective origins, surprises about the identity or attributes of our fossil ancestors may be deeply unsettling and unwelcome. Even professionals, if they are not vigilant, are liable to fall into the trap of refusing to evaluate the evidence objectively; it was precisely this tendency that made it so difficult for Dubois's finds to be accepted.

By the close of the nineteenth century, nearly eighty scientific books and articles had appeared discussing Dubois's *Pithecanthropus,* and hardly any agreed with his interpretation. Worst of all, a German anatomist named Gustav Schwalbe did something that played directly into Dubois's incipient paranoia about intellectual theft: he obtained a cast of the skullcap, courtesy of Dubois, began lecturing about *Pithecanthropus,* and in 1899 published a 225-page monograph on the subject that overshadowed Dubois's more cursory treatments. What a bitter irony that one of the few men to take Dubois's find seriously (Schwalbe thought *Pithecanthropus* was an intermediate between Neandertals and apes) usurped his fossil. After this, Dubois disappeared from the anthropological scene for decades. He stopped attending scientific meetings or publishing about *Pithecanthropus* and refused for many years to let any researchers see the *Pithecanthropus* fossils or the associated fauna. He kept his partial skeleton almost literally in the closet—hidden away inside cabinets in his dining room, with newspaper carefully covering the glass fronts. He was obviously hurt by the hostile reception to his ideas, the antithesis of the triumphal celebration he had every right to expect, and he was increasingly paranoid about other researchers' intentions. He had a mirror over his front door arranged so that he could see callers without opening the door and turn away those who wanted to see his fossils, an arrangement that sounds extraordinary but was common in that region of Holland at the time. There were even rumors he had gone mad and destroyed the bones.

Many people have interpreted this period of withdrawal by Dubois as either avoiding a painful issue or an unbecomingly childish period of sulking. Dubois had certainly suffered harsh treatment and vicious criticism of his theory, and was irritated by others' blindness to what was (to him) the obvious truth. But, to give him his due, Dubois was a strong and stubborn man, not one whose confidence was easily shaken. After all, he had followed his own convictions and defied all common sense in quitting his job and decamping for Indonesia, especially at a time when nearly everyone who thought there *was* a missing link expected it to be in Africa. Was this a man who would turn himself into a scientific hermit over others' opinions? Once I asked John de Vos, curator of the Dubois collection and archives at Leiden, about Dubois's withdrawal. He has a different view of Dubois's later years. "The man was happy," John insisted. "He was working, learning things. He didn't care that no one else agreed with him. *He didn't need their agreement.*" This rings true. If any personality trait of Dubois's is clear after all these years, it is his intellectual independence. Like the Dutch boy with his finger in the dike, Dubois was not about to give up just because his position was a lonely one.

What Dubois was working on would turn out to be almost as important as finding the missing link. He was studying cephalization: the relationship of brain size to body size in various species. He believed that there were rules of cephalization that predicted the ratio of brain size to body weight in any species, according to its place in the natural system, and that these rules showed *Pithecanthropus* to be a perfect intermediate between apes and humans. Whereas previous anthropologists and natural historians had dutifully measured brain size—or, more truthfully, cranial capacity—of any skulls they happened upon, no one else had applied this information to evolutionary history. We know now that the underlying constraint that regulates the ratio of brain size to body size is probably metabolic. The limiting factor on brain size may be the body's efficiency in processing food, for the brain is an expensive organ, physiologically, both to grow and to sustain.

But Dubois was discovering the relationship, not the underlying mechanism behind it. He was searching for an intrinsic law that ruled the relationship of brain size to body size, a magic number or formula that would let him predict the proportions of unknown or extinct animals. From measurements of the skulls of modern mammals Dubois began to perceive the existence of a fascinating series of ratios. He arranged animals in a sequence that roughly approximated that of evolutionary "progress," in reverse, and measured brain size to body size. He regarded the human ratio as perfect and so set it at 1; everything else would be a fraction of this ratio. Apes, the next group of mammals down the scale, had a coefficient of ¼; then came the zoological families of doglike animals (Canidae), cats and their relatives (Felidae), antelope and cattle (Bovidae), and deer (Cervidae), which all showed a coefficient of about ⅛; rabbits had ¹⁄₁₆; mice had ¹⁄₃₂; and moles, bats, and other tiny mammals showed a coefficient of ¹⁄₆₄. This made a regular progression, with each ratio being half of the one that preceded it, with one obvious gap between apes and humans: the missing link, which ought to be ½. When Dubois estimated *Pithecanthropus*'s brain size and estimated its body size, he was able to produce a ratio of ½. It was shaky—he was making two estimates, and the techniques for doing this were only just being developed—but with a few manipulations of the data it worked.

To him, this proved that *Pithecanthropus* was the perfect intermediate between apes and humans. And, as I found out sixty years after Dubois, the ratio of brain size to body size is a key factor in understanding *Homo erectus*. But Dubois's modeling of

brain doubling, from lower to higher mammals, was too sim-
plistic to be correct, and few at the time understood what he was
going on about. He was a pioneer in this area of research and it
absorbed all of his time for years. Then, in the 1920s, *Pithecanthropus* reemerged in Dubois's papers, due in large measure to
pressure from the Royal Dutch Academy of Science. Henry
Fairfield Osborne, the self-important and rather pompous director of the American Museum of Natural History, wanted
casts of *Pithecanthropus* to display in his museum; when these
did not arrive after a year or so, he complained that Dubois was
denying legitimate scientists access to the fossils. It was true that,
over the years, several scholars had written asking to examine
the material, and Dubois had always found it inconvenient to ac-
cede. Besides, he might well have thought, everyone who was
anyone had already seen the fossils in 1895–96 when he took
them all over Europe. Osborne wrote directly to the secretary of
the Royal Academy and demanded that Dubois be forced to
publish the material fully and permit other anthropologists to
see the fossils. Dubois was an employee of the Dutch government. He acquiesced and made arrangements for a London firm
to make and sell casts of his fossils to museums and scientists,
sending him 25 percent of the fees. Still, he grew daily more secretive and suspicious. He resented the fact that no one else
seemed to embrace his theories, though his own opinions were
notoriously inflexible.

3

CHINESE FORTUNES

You might think Dubois's experience would have put off other seekers after the missing link, but it was not so. While Dubois was studying cephalization, Davidson Black, a young Canadian physician, was working with Grafton Elliot Smith in Manchester, England, on human evolution. Elliot Smith, an Australian neuroanatomist and expert on fossil man, was deeply involved in the analysis and interpretation of a fascinating new fossil skull, *Eoanthropus dawsonii*—Dawson's dawn man—better known now as the Piltdown hoax. Dawson was an enthusiastic amateur antiquarian who, by his own or another's design, found pieces of tooth, mandible, and cranium of an apparently ancient human in a gravel pit in Piltdown, Sussex. Most of the material was "discovered" in 1911; an additional tooth turned up in 1913, and quelled the tiny suspicions that were just beginning to germinate. The fossils were, in fact, bits of orangutan and modern human, carefully broken, altered, and stained, then planted among genuine fossils of various animals. It was a cunningly contrived fraud that misled the anthropological community for almost forty years. In 1951, Kenneth Oakley, my friend at the British Museum, was one of the trio of scholars who showed the Piltdown fossils were a fraud.

In 1914, when Black worked in Elliot Smith's laboratory, no one had any thought of a hoax. The fragments of the Piltdown skull were being reassembled and studied for the first time. Although there were numerous quarrels about the exact placement of various fragments and the precise brain size, on the whole the Piltdown specimens seemed to show that large-brained, sapient humans were very old indeed. By implication, small-brained creatures like *Pithecanthropus* could have no place in our direct ancestry. The excitement was tremendous in England. With such a protracted and intense exposure, Black caught a terminal case of fossil fever. He was going to find the missing link for himself. And so, like Dubois, and like me, Black set out for a remote part of the world where he thought fossil hunting might be profitable. Even though Dubois was perceived as having been unsuccessful, scholars in the early twentieth century had come to think that Asia was the most likely birthplace of mankind. (They were wrong, but fashions come and go in science as in everything else.) In 1919, Black took a job at the Peking Union Medical College, teaching neurology, anatomy, and embryology to Chinese students.

Black seems to have had enormous fun in China, though there were unusual obstacles to be overcome. When Black and his wife arrived in Beijing, they were shown to their residence in the university compound: a house that lacked not only furniture but also doors and windows. A colleague who was a professor of anatomy took charge and whisked the dismayed Blacks off on a sightseeing tour of the city, after having a discreet word with someone. When the Blacks returned, the house was fully functional, down to the smiling servant who greeted them at the door. Work proved as challenging, and different, as daily living. When Black attempted to obtain cadavers for gross anatomy dissections, he was told to contact the police, who obligingly sent over a cartload of executed prisoners, all headless. This would not do, Black advised the chief of police, explaining his needs more fully. The next time, the prisoners were sent over intact, and alive, with cheerful instructions that Black was to execute them any way he liked. He didn't like, and worked out other arrangements.

He enjoyed the Chinese and got along well with them, better than most Westerners. He seemed to have a kind of sympathy with their customs and a genuine respect for his colleagues' abilities. Black soon established close ties with Dr. Weng Wenhao, director of the Geological Survey of China, and set about fossil hunting. Fossils were well known in China as "dragon bones," which were believed to have medicinal properties and could be bought in traditional pharmacies. Unanswered were the questions of where the fossils were coming from and whether or not there were ancient hominids among them.

It was the sort of challenge Black enjoyed. Pictures show him as an archetypal fossil jock, handsome and fair-haired, wearing natty jodhpurs and puttees, with round-rimmed glasses perched on his nose and a pipe protruding from his grinning face. He is often posed with his Chinese colleagues, men like Yang Zhongjian, a vertebrate paleontologist from the Geological Survey, or Pei Wenzhong, a bright young man who became the field director of Black's excavations and later earned a Ph.D. in paleontology.

Awkwardly, Black's interest in fossils was seen as a waste of time by the Rockefeller Foundation, which paid his salary at the medical college. They felt strictly medical research should be his occupation. After a visit, one representative of the foundation wrote to Black confessing that he had "jollied you, threatened you, and bullied you, and even cursed you [to give up anthropology], but you took it well. . . . For the next two years at least give your entire attention to anatomy. Perhaps by that time, you will, with your young son, have other interests which will appear more important than expeditions to mythological caves." The foundation was so against Black's anthropological endeavors that it withdrew funding for a series of lectures after discovering that (at the request of Black and his co-workers) they were to be given by Aleš Hrdlička, the physical anthropologist and "fossil man" man at the Smithsonian Institution, because he was sure to encourage Black's interest in caves and fossils. This was such a public embarrassment to both Black and Hrdlička that the foundation relented somewhat. They still refused to pay Hrdlička's way but, deviously, made a grant to the Smithsonian on the understanding that it would be used for his fare.

Nonetheless, Black wouldn't stay away from the mythological caves, especially one close to Beijing called Zhoukoudian (spelled, at the time, Chou Kou Tien), or Dragon Bone Hill. A few years after Black's arrival in China, Swedish scientists working with the Geological Survey of China found numerous fossil bones there, along with oddly shaped bits of quartz that they thought might be ancient stone tools, and two isolated teeth, both in rather bad shape. Black and J. Gunnar Andersson, of the Swedish team, were convinced they were human. They laid elaborate plans for a thorough excavation of Zhoukoudian and used the visit of the Swedish crown prince (later King Gustavus VI) to Beijing in 1926 as an opportunity to win funding. The prince was an amateur archaeologist and the Protector of the Swedish Research Committee, which controlled all funds for fossil collecting abroad. At Andersson's request, Black prepared a report about the teeth to be read at a meeting with the prince.

Black knew how to sell a project, even if two isolated teeth were all he had to go on. He described them as "a striking confirmation" of the prediction that China would yield a "new Tertiary man or ancient Pleistocene man" that was linked to *Pithecanthropus* in Java. "The Chou Kou Tien discovery therefore furnishes one more link in the already strong chain of evidence supporting the hypothesis of the Central Asiatic origin of the *Hominidae*," Black concluded. Andersson added that it might be the most important achievement of all of the Swedish archaeological endeavors in China. (At the time, archaeology, the study of stone tools and other objects made by humans, was commonly muddled up with paleontology, the study of the bony remains of humans and animals.) The prince was won over.

The prince's backing caused the Rockefeller Foundation and the Peking Union Medical College to reverse their opinions of fossil man research. The foundation provided long-term financial support for systematic excavation at Zhoukoudian and for setting up an Institute of Human Biology and a Cenozoic Research Laboratory at the college, which would carry out the research and laboratory work and house the fossils and skeletal materials.

Perhaps Black thought that, with funding and approval in hand, finding the missing link would be a snap. Fieldwork began in earnest on March 27, 1927, with the researchers and technicians housed at the only local accommodation, a camel caravansary called the Liu Shen Inn. Black supervised, but left the day-to-day work to Anders Birger Bohlin, a Swedish paleontologist, or Li Jie or Yang Zhongjian, Chinese geologists, not to mention sixty laborers. By the end of the first season, the team had found another hominid tooth. Black was ecstatic. "We have got a beautiful *human* tooth at last!" he wrote to Andersson in Sweden. "It is truly glorious news, is it not!"

Black had fewer fossils than Dubois—only one that his expedition had found, and all of them could get lost in a handkerchief—but he was no less convinced. He rushed into print, naming a whole new genus and species on the strength of those few precious teeth. It was *Sinanthropus pekinensis,* honoring China (*Sin-*), mankind (*anthropus*), and Beijing (*pekinensis*). (It was, of course, simply *Homo erectus,* the same species that Dubois in Java had called *Pithecanthropus erectus,* but Black didn't know that.) Black promptly left for a tour of Canada, the U.S., and England, where he displayed his fossil tooth, which he carried in a special brass receptacle worn either on his watch chain or around his neck. Black argued his rather thin case for a new genus and species and pleaded with the Rockefeller Foundation for more money. Black won on all counts, even if he was regarded as having rather hastily conceived a new genus and species. He knew, better than Dubois, how to seek the support of those influential in his science.

In 1929 he was vindicated. By then Pei Wenzhong was in charge of the excavations at Zhoukoudian. Though he was less experienced than the joint team he replaced and was perhaps even a little intimidated by the responsibility, he had been trained on the job and knew his stuff. (Black once called him "a corking field man," and so he was.) On December 2, working in cramped conditions in a tiny part of the cave, Pei excavated a hominid skullcap by candlelight. It was more complete than Dubois's and went from the browridges to the back of the skull and around to the preserved underside; only the face was broken away. Pei knew his first duty was to tell the expedition's leader, as when Kamoya called Richard and me on the radiophone. "Found skullcap—perfect—look[s] like man's," Pei telegraphed Black. He knew it was the most important thing they had found yet, so he set about preparing this delicate find for travel.

He had removed the skullcap from heavy clays; it was wet and soft, a dangerous state. For the next few days, Pei and his helpers took turns watching the fossil by the fireside, carefully turning the specimen as it dried slowly. Pei wrapped it in glue-soaked gauze, covered it in plaster, then wrapped the whole mess in two thick cotton quilts and two blankets tied on with rope, and set out for Beijing. Ahead lay a bus ride and, ominously, a police checkpoint where all luggage had to be examined. Pei feared difficulties and readied himself to defy authority if necessary; the fossil had to be kept safe. He later reminisced about his anxieties on that trip:

> I made preparations for . . . [the police check] with a few fossils to show the officer, intending to tell him it was the same kind of thing inside the wrapped luggage and to ask him not to open it. If he insisted on opening the bundle, the plaster and gauze would have to be kept intact. If he still insisted on taking a look at what was inside, I would ask him to arrest me first. [Fortunately] the man was polite. I only had to open the bundle to let him see.

By December 6, the fossil was in Black's hands at the new Cenozoic Research Laboratory, and Black knew Pei's enthusiastic telegram was no exaggeration. He cleaned the skull, using a needle to pry off each bit of clay, and glued a few cracks and breaks. With advance notice to Elliot Smith, his old supervisor, and a few other influential men, a beaming Black announced the find to a special meeting of the Geological Survey of China and the press on December 28, 1929. His colleagues were present to share in the glory. Pei Wenzhong described the circumstances of the find, Yang Zhongjian spoke of the geological setting, and Pierre Teilhard de Chardin, a French priest and paleontologist attached to the Geological Survey, provided information about the fauna associated with the skull. Weng Wenhao, proud director of the Geological Survey, presided over the meeting. Their excitement was infectious, and the journalists who were present made headlines around the world with the announcement.

Black knew how to spread the news effectively. He embarked on a European tour, with casts and slides and a well-polished lecture that impressed all who heard him, unlike Dubois, whose tour had improved his standing but couldn't dig him out of the hole of skepticism into which he had fallen. Black never risked disapprobation. For one thing, he wasn't facing Virchow and the "there's no such thing as evolution" school; for another, Black's tactful advance warning to Elliot Smith helped pave the way for his celebratory reception long before his arrival.

Black had found the missing link, but the name *Sinanthropus pekinensis* persisted for years before paleoanthropologists were convinced that it and Dubois's *Pithecanthropus erectus* were the same. Black suggested that *Sinanthropus* was an intermediate between *Pithecanthropus* and Neandertals. Dubois himself was never persuaded; he asserted vigorously that his beautiful *Pithecanthropus* was the missing link and in no way closely related to Black's *Sinanthropus*, which was perhaps a Neandertal, basically human. Very late in his life, Dubois even began to hint that *Pithecanthropus* was a giant gibbon that had somehow been ancestral to humans, an awful replay of Virchow's early accusations. Dubois died in 1940 and on his grave is carved a skull and crossbones, which represent the likeness of the missing link that usurped his life.

Black came to what was probably a happier end. Once the first skull was found, he was hailed by the international scientific community as the finder of the missing link. In the years that followed, the excavations at Zhoukoudian continued to be rewarding. Though the first skull proved to be that of an adolescent, a second, adult specimen was found, as well as a mandible, or jaw. None was complete, and unfortunately there were no faces. In 1932 Black was elected a Fellow of the Royal Society for his efforts. The team also found many stone tools, evidence of fire, and lots of fauna, many with broken bones suggesting they might have been food refuse. Jia Lanpo, a young archaeologist, was brought in to study the stone tools and deduce how early hominids lived in China.

Excavation style at Zhoukoudian in those days was radically different from ours now. Black always opened and closed the field seasons himself, but he left much of the operation in the hands of the field directors. At first, the laborers dug at random until they found a fossil or stone tool, at which point they would call over a technician or a supervisor to remove it. Being no fools, the workmen dug deeper where the matrix was soft and left the difficult places standing up like pedestals. This meant that the sequential numbers assigned to fossils or tools bore no relationship to where the object was found, nor did any other data. In 1932 the Zhoukoudian team revised their excavation methods to provide more systematic control over the in situ location of each find. This was an important step, since there were now an unbelievable one hundred to two hundred people digging at two or three individual sites that were spread over two square kilometers. With so many workmen, there was no way Black, Pei, or any of the other supervisors could remember each individual fossil. The new procedure was to dig in trenches measured into one-meter blocks, each of which was numbered, and to give a number to each specimen that began with the block number. A grid was sometimes physically painted onto the rock, along with the block identifiers. (Numbers were assigned to geological layers descending from top to bottom; east-west, blocks were given letters, and north-south they were assigned numbers, so a particular fossil's location might be layer 6, block 13F.) The supervisors took photographs from standard points at the beginning and end of each day, which is not as exact as our method of measuring each fossil's position in three dimensions but was a landmark step in those days.

The scale of excavation was enormous, a nightmare. Backdirt was hauled off in baskets carried on shoulder poles by a steady stream of laborers. In 1933 Pei coined the slogan "Mechanize the Digging," and speeded up operations considerably by setting up a clever cable-and-pulley system. Full baskets slid downhill along a cable, moved by their own weight, to the workers below, who waited to empty them; their movement propelled the empty baskets uphill along the second stretch of cable. No wheelbarrows or *kerai*-loads for them.

Black usually stayed in Beijing, working through the night on the hominid and mammalian fossils at the Cenozoic Laboratory. He did much of the preparation of the fossils with a dental drill. He also liked to do the delicate gluing and reconstruction; like me, he must have found it a very satisfying task. He preferred to work uninterrupted in the empty building, returning home in the early hours of the morning to sleep until noon. But he had a congenital heart condition, and the strain of his schedule began to tell. In February 1934 he was hospitalized for six weeks; once released from the hospital, he started in again enthusiastically. His enthusiasm was perhaps too great. He was found dead at his desk by his colleagues on the morning of March 15, not yet fifty years old. He was so beloved that, years later, on the anniversary of his death, the entire Department of Anatomy and the staff of the Cenozoic Laboratory still went out to the European cemetery in Beijing to leave flowers on his grave.

Dubois found the missing link, but sacrificed his scientific credibility to it. Black found the link and gave up his life. With the growing sense of continuity that connected their lives to mine, I couldn't help wonder what the missing link would bring to me. The risks are clear, even for those with the good luck to find great fossils. They may bring acclaim or scathing disregard, fame and fortune or illness and death. One thing is for sure, though. Once you've found an extraordinary fossil, you will not suffer from boredom.

4

THE MAN WHO LOST
THE MISSING LINK

So many people seem to think that by studying the past they can understand the present, or even predict the future. I'm not so sure. I have always been struck by L. P. Hartley's observation: "The past is a foreign country; they do things differently there." This sentiment surely applies to the remote past, when *Homo erectus* lived. But through my involvement with the Nariokotome boy, I became a member in a special sort of club comprised of those who have been captured by the missing link. Their findings and ideas were the foundation upon which my work would be built. I couldn't hope to understand the boy in my own terms unless I could comprehend what had been done with earlier finds of *erectus*. There are lessons, too, in my predecessors' mistakes and in their foibles, morals that might apply as much to my story as to theirs. I found myself fascinated by the lives of the other members of the club, like Dubois, Black, and Black's successor, Franz Weidenreich.

Black's death didn't stop the excavations at Zhoukoudian. The field season at the prime hominid locality was up and running again a month later, in April 1934. It may seem heartless, but it is not, for finding the elusive answers to the questions of our ancestry drives us all. When a colleague dies, the work becomes a sort of tribute to his contributions.

Black's successor, Franz Weidenreich, would become the man who lost the missing link. Weidenreich was a German anatomist who was a well-known expert in fossil man, trained by Gustav Schwalbe, the anatomist who had appropriated Dubois's *Pithecanthropus*. In 1934 Weidenreich was forced to leave his professorship in Frankfurt, at the age of sixty-one, because he was Jewish. All his publications written before leaving Germany, with one exception, were in German; after that, none was. He took American citizenship and was teaching at the University of Chicago when he was offered the chance to go to the Peking Union Medical College to replace Black. In doing so, Weidenreich took over one of the most important fossil man digs of that era, which carried with it the privilege and responsibility of formally describing Peking Man, as the fossils were then called.

It was uncomfortable stepping into the shoes of one as beloved as Black. Unlike his predecessor, Weidenreich was no fossil jock, and though courtly and correct he never had the easy rapport with the Chinese that Black had enjoyed. Weidenreich was small and precise, taciturn, formal, and a bit fussy, covering his bald pate with a beret and wearing a tie and a three-piece suit even at the site. The Chinese sensed that he visited the excavations in part to ensure that the work was being done carefully. He even admitted that, at first, he feared that the Chinese were carelessly discarding the limb bones of *Homo erectus*—why else were only skull bones being found?—though he revised his opinion after he saw their procedures and techniques for himself. The Chinese, however, never forgot his initial qualms about their competency.

Still, he was a superb anatomist and analyst. Describing fossils anatomically is an exacting task. You must put together a systematic, thorough word-portrait of the size and shape of each bump, lump, and crest on a fossilized bone, and compare it with all known closely related fossils. Some people never develop the gift for writing descriptions. It requires a precision of language and an exhaustive knowledge of the anatomy of the region of the skeleton you are describing. Working through someone else's description while you hold a fossil (or a cast of it) in your hand gives you a clear glimpse of that person's competency and turn of mind; you discover what someone else saw in the fossil as well as how carefully he marshaled his observations. This is the only way I have come to know Weidenreich. I have noticed that his descriptions are so thorough that he defined new terms for previously unremarked anatomical structures. His meticulous scholarship earned the respect of his Chinese colleagues, as it earned mine, years later, when I had to consult his works closely. He was, by all accounts, a kindly man too; his secretary, Claire Taschdjian, was reportedly devoted to him for his considerate manner and his willingness to teach her about the fossils.

The other side of his precise nature was his penny-pinching. In his memoirs, Jia Lanpo, who for years supervised the fieldwork, talks about the difficulty of complying with the directives of the "old gentleman," as Weidenreich was called, and still getting the work done. Sometimes the Chinese even plotted behind Weidenreich's back. In 1936, his compatriot, the geologist Bian Meinian, wrote to Jia:

> The Cenozoic Laboratory has received the grant from the Rockefeller Foundation for this six months' expenditure, but the old gentleman is so possessive that he would appropriate

only one thousand dollars for the fieldwork each month. You need not worry, however, if you have to spend more, say, two, three or five hundred more. I'll fight the battle at the front for you. The old gentleman has a notion that the fieldwork has to be done slowly and cautiously, oblivious of the fact that the piled-up soil has to be sifted, and that costs money. You should just go ahead and have the soil sifted. . . . Weidenreich is so preoccupied with controlling expenditures that he gives no thought to the fact that money unspent has to be handed back.

Despite Weidenreich's financial anxieties, the site yielded ample material—some five hundred boxes of specimens in two field seasons. But there were no homs until October 1936, when the Zhoukoudian crew found a good hominid mandible, followed shortly by the most exciting find in years: a new skullcap. It was in pieces, but Jia spotted it as a technician was placing the first fragment (which he did not know was a hominid) into his wicker basket. By the end of the day, enough pieces had been found to show it was another partial skull: Jia sent word to Weidenreich in Beijing to come quickly. (Weidenreich was so excited when the telephone message came the next morning that the normally imperturbable scholar put his trousers on inside out.) This skull, when glued together, was the best preserved of the three found at Zhoukoudian, though the face was still missing.

Weidenreich's newborn good luck was short-lived, slain by world politics. The subsequent field season was disrupted by the Japanese invasion of China, which came close to Zhoukoudian. In 1937 only a "skeleton crew" continued working at the site; in early 1938 the site was left to a few caretakers who were local residents. The Japanese started expressing considerable interest, photographing and measuring the excavations; against orders, the caretakers refused to abandon their precious site to these interlopers. In May, three of the crew were captured and tortured to death by the Japanese, who believed the men knew the whereabouts of a band of guerrilla fighters. The fossils were also in danger from the Japanese, who showed increasing interest in the Peking Man specimens throughout 1940 and 1941. Many Americans fled China during these years, for safety, and Weidenreich made plans to return to the United States in the autumn of 1941, where he had been offered refuge as a visiting scholar at the American Museum of Natural History in New York. The Rockefeller Foundation agreed to continue to pay his salary as a professor of the Peking Union Medical School, even though it was too dangerous for him to be in Beijing.

Though he had to leave his precious fossils behind, Weidenreich was determined to take as much information with him as possible. The thought that the Japanese were planning to loot the Cenozoic Laboratory was on everyone's mind, especially because two professors from the Tokyo Imperial University seemed to be camped in Beijing, awaiting a chance to examine—or appropriate?—the fossils. Weidenreich asked his assistant, Hu Chengzhi, to make exacting plaster casts of every hominid fossil, not just the "good" bits. Weidenreich himself embarked on the nit-picking task of photographing, drawing, and measuring each piece and taking detailed notes that would let him continue his anatomical studies in New York. It was a daunting task for both men; there were dozens of specimens, all priceless and many fragile.

Conditions grew so tense that Weng Wenhao, the director of the Geological Survey, asked Weidenreich to take the fossils out with him when he fled China, which placed Weidenreich in an awkward position. He wanted to protect the fossils and doubtless wanted to oblige Weng, but he was no longer a young man and had perhaps never been a daring one. Besides, he had already been chased out of Europe and one homeland by the Axis powers. As a well-known scientist sympathetic to the Americans, it was likely that his luggage would be searched. If the fossils were seized, he would be responsible for their loss and might well be imprisoned for attempting to remove them from China. He had not even the slimmest legal pretext for exporting them, since the detailed agreement that Black and Weng had drawn up at the outset of the Zhoukoudian project stipulated that the entire collection of fossil and archaeological specimens was to remain in China. Weidenreich reluctantly refused the request.

Weng decided that the fossils had to be moved from Beijing, so, even before Weidenreich left, Hu and one of the other technicians packed up the fossils. According to their memoirs, each fossil was wrapped in white tissue paper, cushioned with cotton and gauze, and then wrapped again in stouter paper. These packages were placed in a small wooden box, which was in turn padded and placed with other boxes inside large wooden crates. Weng appealed directly and successfully to the American ambassador to get the fossils out. The boxes were delivered to the U.S. Embassy in November 1941, where they were turned over to Colonel William Ashurst, who was instructed to include them with his personal belongings and treat them as secret materials. A contingent of nine marines was to shepherd the boxes on a train journey to Qinhuangdao (then Chinwangtao), on the coast, where they would be picked up by the SS *President Harrison*. The train left Beijing at 5:00 a.m. on December 5, 1941, and arrived at the coast on December 7; the fossils and the marines were to sail the next day.

December 7 was the "date which will live in infamy," in Franklin D. Roosevelt's words: the day the Japanese bombed Pearl Harbor and brought the United States into the war. The plan for the fossils' evacuation fell apart. The train carrying the marines was intercepted by Japanese troops en route. Besides, the *President Harrison* never made it to Qinhuangdao; she was scuttled by her crew in a fruitless attempt to keep her out of Japanese hands. The marines were sent to prison camps, their ammunition and military equipment was seized, and the fossils were never seen again.

Had the fossils left Beijing? Had they reached the coast ahead of the marines, as had much of their other gear, or were they on the train? Were they languishing, unrecognized, in a warehouse somewhere? To this day, no one knows.

Apparently, the Japanese did not capture the fossils, for they instituted earnest and repeated searches for them in Beijing shortly afterward. The two Japanese professors appeared at the medical college, frustrated to find that the safe that had once contained the fossils now housed only casts. But they took to Tokyo the stone tools, the catalogues, and some of the mammalian fossils from Zhoukoudian that had been left behind in the lab. For the next three years Japanese officials persisted in questioning and intimidating employees of the Cenozoic Laboratory, especially Pei Wenzhong. No one was spared. Weidenreich's secretary, Claire Taschdjian, was arrested and escorted through warehouses under military guard, to look for the missing boxes. The manager of the medical college, Trevor Bowen, was questioned and imprisoned for five days in a wooden cage so small he could not lie down, until the Japanese were convinced he knew nothing of the fossils' whereabouts. Even the American embassy was searched, with no success.

The missing link was lost again.

What must Weidenreich have felt like? If it had been me, I would have been devastated. Under Weidenreich's guidance the collection had grown from Black's two skulls, one mandible, a few teeth, and other fragments to boxes of specimens that represented at least fifteen individual *erectuses*. All were carefully documented, thanks to Weidenreich's foresight and to Jia Lanpo, who had secretly copied all the field notebooks onto tissue paper and smuggled them out of the Cenozoic Laboratory under the noses of the Japanese soldiers. In New York, Weidenreich learned of the fossils' fate indirectly. His Chinese colleagues felt it dangerous to communicate with him too openly, but one managed to get him a postcard that read simply,

"Where's Nelly?"—using the lab's nickname for a fleshed-out reconstruction of one of the skulls. Weidenreich surmised the awful truth.

Rumors flew concerning the fate of the fossils for years, but no one knew where they were. After the end of World War II, Weidenreich arranged to bring to the American Museum of Natural History another displaced person, his Dutch colleague G. H. R. von Koenigswald, known as "Ralph" or "von K," who before the war had worked for the Geological Survey of Java and had revisited some of Dubois's sites as well as Solo, another site near where the first *erectus* had been found. Weidenreich had visited von K in Java and had collaborated with him on describing and analyzing the skulls from Solo.

In 1945, American troops under the command of Frank Whitmore, Jr., of the Military Geological Unit of the Allied Headquarters, searched specifically for the missing fossils in Tokyo. This led to the recovery of the stone tools, some mammalian fossils, and many of the catalogues and record books from Zhoukoudian, but not the Peking Man fossils. Remarkably, however, one of von K's skulls *did* turn up. It was called Solo IX (like the Zhoukoudian fossils, the fossils from Solo were assigned roman numerals for identification). The skull was recognized by Walter Fairservis, an army lieutenant (and future archaeologist) who happened to be part of the group that occupied the Imperial Museum in Tokyo. Solo IX had been sent to the Japanese emperor as a birthday gift. Fairservis simply brought the skull to the American Museum of Natural History and, a few days before Christmas, presented it to an astonished von K, who later took it with him to Holland and then much later to Germany. (Although the Solo skulls are now back home in Java, many of von K's fossils were his personal possessions and did not belong to either the Dutch or Indonesian governments. Many still reside in the Senckenberg Museum in Frankfurt, the last institution at which von K worked.) Hoping in vain for another miracle, Weidenreich lobbied to get Fairservis assigned to search for the Zhoukoudian fossils too.

The return of Solo IX led to what must have been Weidenreich's bitterest moment. Soon after Fairservis brought the skull to New York, a British vertebrate paleontologist, D. M. S. Watson, was visiting the American Museum of Natural History. Weidenreich showed him the Solo skull as well as the casts of the Zhoukoudian material. Being no paleoanthropologist, Watson apparently confused Solo IX with the Zhoukoudian IX skull, which he knew was "missing." Over the next year or so, Watson inadvertently misrepresented what he had seen when describing his New York visit to a number of colleagues, including the East German paleontologist Walter Kuhne. Kuhne concluded that the deceitful Americans had had the Zhoukoudian fossils throughout the war, and in December 1951 he informed Black's old associate Yang Zhongjian, who was by then the director of the Institute of Vertebrate Paleontology and Paleoanthropology of the Chinese Academy of Sciences, of this false conclusion. Both Yang and Pei Wenzhong, who had had major responsibilities at Zhoukoudian, wrote indignant letters demanding the return of the fossils, incensed by the apparent perfidy and duplicity of a man they had thought their friend and colleague.

For his part, Weidenreich was hurt that his former colleagues could believe such a thing of him, without (as far as he knew) any shred of evidence. Harry Shapiro, then curator of physical anthropology at the museum and Weidenreich's boss, was deeply embarrassed by the accusation; he issued a flat denial in writing and through the press. Weidenreich could do nothing but continue to work, publishing scientific monographs and papers about the Zhoukoudian fossils until the day he died in 1948.

First, he recognized that Dubois's *Pithecanthropus* from Java was little more than a local version of *Sinanthropus* from China. He argued convincingly that they differed only in ways rather akin to the racial variations found among humans from those regions of the world. Yet he clung stubbornly to the practice of calling the specimens by different Latin names. By 1946, Weidenreich knew that modern evolutionary biologists were emphasizing what was called the *biological species concept*. For two individuals to be put into the same species meant that they were literally (or potentially) able to interbreed—as are, for example, a Chinese woman and a Javanese man. To put two individuals into *different* species (or, more dramatically, into different genera) implied a profound barrier to interbreeding and markedly different adaptations. These are the standard assumptions that now underlie modern taxonomy, or classification.

"However," Weidenreich wrote, "I have refused [to comply with these procedures]. In paleontology it always was and still is the custom to give generic or specific names to each new type without much concern for the kind of relationship to other types formerly known. Furthermore, the old names of fossil human types are accepted throughout the entire literature dealing with early man, so that any radical change would lead to the greatest confusion and necessitate complicated explanations in each case. I shall continue, therefore, to use the old names without imputing a special taxonomical meaning." He was tartly critical of other scientists who endangered anthropological progress by "prejudice and obstinate retention of out-moded ideas," but he remained unwilling to bring paleoanthropological naming practices into line with modern biology. To be fair, it took almost fifteen years before most anthropologists gave up these older names and placed all the specimens from Java and China (and elsewhere) into *Homo erectus*.

Weidenreich also realized that one of the main features of the evolutionary process that changed *Homo erectus* into modern humans was the enlargement of the brain. He carefully documented the steady enlargement of the cranial capacity of the skull over time. The Javan specimens (which he believed were older) had about 860 cc of braincase. This went up to about 1,075 cc in the

Zhoukoudian fossils, and rose to about 1,400 in Neandertals, whose brains were about the size of modern human brains.

Even though he called *Sinanthropus* and *Pithecanthropus* by different names, he was quite clear that they were the same sort of hominid and that they were part of an evolving lineage that led to modern humans. His studies of these fossils led him to propose an important theory, one that was not widely accepted at the time but that has proven remarkably persistent and influential. Weidenreich's regional hypothesis traced various anatomical features found in ancient populations to similar characteristics of the modern groups that later inhabited the same region. For example, he noticed that both *Sinanthropus* and living peoples of Asia had an unusually high proportion of shovel-shaped incisors, or front teeth, in their upper jaws. (By "shovel-shaped," Weidenreich meant that the back or tongue-side surface of the teeth had a ridge along the gumline and down each side, making a shape like the blade of a shovel.)

To explain his hypothesis, he drew ungainly diagrams that showed *Pithecanthropus* evolving into modern Australoid peoples and *Sinanthropus* evolving into modern Mongoloid peoples. This evolutionary pattern was symbolically represented as longitudinal parallel lines, each decorated with fossil names and each terminating in a living human race. He also drew confusing crisscrossing lines, at right angles and at forty-five degrees, which were supposed to indicate genetic continuity (interbreeding) among the separate lineages. Gene flow among different populations kept them all part of one vast, polymorphic human species. The overall effect was like a diagram of a very complex play in American football; you had to be in on the drawing of it to understand. This diagram was later simplified and transformed by Harvard anthropologist William W. Howells into the "candelabra hypothesis." In a much-read book published in 1959, *Mankind in the Making,* Howells showed Weidenreich's hypothesis as a candelabra, with each arm or each geographic lineage coming from a common origin and ending in a living race. The gene flow among lineages, which might have been portrayed as a spiderweb draped across the candelabra, was omitted in the interests of simplicity. Those who failed to go back and read Weidenreich's original works were misled by this diagram into thinking Weidenreich foolishly believed there was no interbreeding or genetic contact among the races since *Homo erectus* times.

If, as Weidenreich proposed, these early hominids had "special affinities" with the modern groups living in the same areas, then the regional differentiation of humans into the locally adapted groups we call races is a very ancient process indeed. Weidenreich's idea depended on a careful inspection of tiny anatomical details: the shape of the upper incisor teeth; the way the sutures join one bone of the skull to another; the presence or absence of a slight ridge or keel that runs from front to back down the center of the skull; particular, characteristic bumps and lumps on the mandible; and the shape of the shaft of the femur in cross section. It was frustrating for Weidenreich that few paid much attention to his regional hypothesis during the last few years of his life. His anatomical observations are sound, but their significance, and his theory (now revived by a new generation), is still the cause of fervent arguments at paleoanthropology meetings, where the origin of modern humans and the antiquity of racial differences is much debated.

The mystery of the loss of the Zhoukoudian fossils continued long after Weidenreich's death in 1948, with increasingly Byzantine complications. In 1971 the entire affair surfaced again because a physician—an ex-marine—William T. Foley, suddenly announced that he had been the last person entrusted with the Zhoukoudian fossils. While stationed in China, he said, he had been put in charge of them by Colonel Ashurst. The story came to light because another ex-marine from China who now worked for Foley, Herman Davis, knew that Walter Fairservis had found the Solo skull in Tokyo. Davis telephoned Fairservis on Foley's behalf to ask if the Peking Man fossils had ever turned up, and Fairservis, convinced that Foley knew something about the fossils' whereabouts, contacted Harry Shapiro at the American Museum of Natural History.

Shapiro promptly interviewed Foley and Davis, who had been a pharmacist's mate stationed at Camp Holcomb in Qinhuangdao in the autumn of 1941. The story they told was tantalizing. Foley had been in command of a unit of seventeen marines in nearby Tientsin. They, like all other marines in the area, were to be transferred to the Philippines on board the S.S. *President Harrison,* which was due to leave Qinhuangdao on December 8. Davis told Shapiro that he had been warned that some special footlockers marked with Foley's name would arrive from Beijing by freight train; these would contain personal effects and "secret material" and were to be carefully guarded until the December 8 departure. The footlockers arrived on December 7 (which was still December 6 across the date line in America), and Davis stored them in his own room for safekeeping. When he awoke on the morning of December 8, the Japanese had bombed Pearl Harbor and Camp Holcomb was surrounded with Japanese troops. Davis surrendered as ordered. Each of the Americans was permitted to take a single bag of belongings to the prison camp at Tientsin, the rest of their belongings to be inspected and forwarded to them.

As foreign military officers, Foley and Ashurst were also arrested on December 8, but for some reason Foley was placed on a sort of extended house arrest for about a week. He was allowed to remain in his home in the British Concession and to move around Beijing attending to patients and hospital duties. Anticipating that he would, in the end, be imprisoned, Foley began to hide the footlockers, presumably including some of the ones containing the fossils, at various locations for safety: at the Swiss Warehouse, at the Pasteur Institute, with Chinese friends.

At this point, disconcerting anomalies appeared in the so-far plausible tale. Foley was never intercepted with the footlockers or questioned by the Japanese about his odd behavior. And he remembered having four or five footlockers, rather than the two that were packed at the Cenozoic Laboratory. Foley also recalled that the fossils were packed in large glass jars within the footlockers, whereas Hu Chengzhi, Weidenreich's assistant, had used boxes. Had the fossils been repacked? Why? (Glass jars would have been more of a danger than a protection for long-distance transport.) And by whom? After a few days, Foley and Ashurst were sent to join Davis and the other men at the prison camp. For years, through various maneuvers and subterfuges, they managed to hold on to one of the footlockers containing fossils, without its being inspected. But it eventually disappeared when Foley and Ashurst were split up and sent to different camps.

While various points in the story seemed incredible to Shapiro, much of it was plausible. Before Shapiro could organize a search of the locations where Foley said he had deposited various footlockers—it would prove fruitless anyway—another searcher appeared and the story got even stranger. The newcomer was Christopher G. Janus, a stockbroker from Chicago, who had just returned from a trip to China in 1972. He was one of the first Americans allowed into China in years and had gone hoping to set up contacts for cultural tours and exchanges to be run through an organization he headed, the Greek Heritage Foundation.

One wonders what the attraction of the missing link was for Janus. Fossil hominids have never made anyone rich—he seems to have been an entrepreneur through and through—and he had no idea what the Zhoukoudian fossils represented to science. According to a popular book Janus wrote, he was drawn into the mystery by happenstance. During his tour of China he was shown around Zhoukoudian and through the Peking Man

Museum; Janus claims that Wu Ching-chih, the director (and a paleontologist), took him aside personally and begged him to try to find and return the fossils. Why Janus's aid would be enlisted is unclear. Janus seemed to have connections with many people in high diplomatic and government circles, such as then Secretary of State Henry Kissinger. Perhaps Wu felt Janus was not the innocent tourist he claimed to be but a quasi–government official. In any case, Janus writes that this request for help was repeated or referred to obliquely several more times during his visit and, upon leaving China, he was met by a group of reporters who inexplicably asked him about the fossils. He responded by announcing a $5,000 prize for information leading to the discovery of their whereabouts.

He contacted Harry Shapiro at the American Museum, hoping Shapiro would agree to authenticate any material that turned up. Shapiro disliked the idea of a reward, feeling the publicity would attract cranks and charlatans. He felt justified when Janus received hundreds of strange claims. One of Janus's "best" leads was a call from an anxious woman who refused to give her name. Her late husband had been a marine in a prison camp in China, she claimed, and had brought home a footlocker full of bones. He had told her they were extremely important and valuable but that she might be in danger if anyone knew she had them. She seemed to fear criminal prosecution and perhaps even physical attack. After several aborted appointments, she agreed to meet Janus on the observation deck of the Empire State Building at noon. She brought a poor photograph that seemed to show a litter of bones lying in the bottom of a box. She was extremely nervous and fled when some tourists approached with cameras in hand.

When she called Janus again, she asked for a half-million-dollar ransom for the fossils, as well as legal assurances that she would not be prosecuted. Janus pressed for another meeting with the woman so Shapiro could examine the bones or at least see a copy of the photo; he also suggested dealing through an intermediary, a lawyer, to protect her from whatever she feared. A series of rather paranoid transactions ensued. Somehow Janus managed to extract two letters from the U.S. State Department: one from John Richardson, Jr., Assistant Secretary for Educational and Cultural Affairs, asserting that the mysterious lady was in no danger of prosecution; the other from Francis B. Tenny, director of the Office of East Asian and Pacific Programs, avowing that the U.S. government had no interest in confiscating the fossils should they turn out to be genuine. Both letters mentioned that it would be good for Chinese-U.S. relations if the fossils were found and returned. Eventually Janus obtained a copy of the photograph, but the mysterious lady disappeared and could not be traced, even by the FBI and the CIA.

The photograph is an enigma. Most of the bones in it look like modern human bones. Many of the skeletal elements pictured are unknown from Zhoukoudian. Both Shapiro and another paleoanthropologist who examined the image fixed on a light-colored shape in the upper-right-hand corner of the photo that *might* be one of the Zhoukoudian skullcaps. Looking at the photo today, I have to say that this *could be* an image of one of the Zhoukoudian skulls, or of a cast of it. But there is a strong whiff of fraud or flimflam about the entire episode. In the years that followed, Janus raised more money, offered a larger reward, and pursued one wild goose after another. Finally, in 1981, Janus pleaded guilty to two counts of an indictment charging misuse of $640,000 he had raised to search for the missing fossils.

The Empire State woman, or whoever took the photograph, may have been earnest but mistaken. The general public has only a very hazy idea of what fossil human bones look like. An enthusiast once turned up at my office with a huge stone that filled the trunk of his aging Chevy, the specimen so heavy that the car's front wheels were almost off the ground. It was, he told me solemnly, a fossil human skull; see? there's the eye, the nose, the ear. . . . When I asked him to compare its size to that of his own head, he looked puzzled, then disappointed. So I think that perhaps this woman believed her husband had brought back some important old bones, but they aren't likely to have been those from Zhoukoudian.

Peking Man—Weidenreich's lost link—is still missing. All we have to go on are Weidenreich's notes, photographs, drawings, and casts. In recent decades, Jia Lanpo, Pei Wenzhong, and other colleagues have been restored to positions of authority. Excavations at Zhoukoudian have been reopened and a few new pieces have been found that would clearly fit onto the original skulls, if only *they* could be found. I hope they rest somewhere, still intact, where they will be stumbled across and recognized. Museum and university basements and warehouses often house strange heavy boxes laden with dust that no one remembers the origins of and that no one troubles to open for years on end. Once I opened a dusty jar in a storeroom at Johns Hopkins Medical School only to find an embalmed gorilla face staring back at me. Maybe one day someone will open another receptacle and gaze into the eye sockets of Peking Man once again.

5

MISSING LINKS FROM THE MISSING CONTINENT

It is of more than simple historical interest to me to learn what those who found *Homo erectus* before me thought about that species. Rereading their works helps me see both what has been established on firm ground and what has always been assumed to be true but actually rests on shifting sands. The distinction can be crucial. For example, Weidenreich may have been cursed to be the loser of the missing link, but he also sketched for us many of the anatomical features that show the evolutionary continuity between *Homo erectus* and modern humans. Some would say, in fact, he drew the connections too finely, seeing racial variations and continuity that stretched back hundreds of thousands of years. In Weidenreich's time, anthropologists' ideas of how long ago *Homo erectus* lived were a bit shaky, but they often guessed at about half a million years.

As is usual with such ideas, new finds and new techniques overturned them. In the early 1960s came the application of radiometric dating, the field that has rightly been called the "oldest" profession, to hominid evolution. Radiometric dating is based on the steady rate of decay of radioactive isotopes into stable, daughter isotopes, such as the transformation of radioactive potassium-40 (^{40}K) to stable argon-40 (^{40}Ar). When volcanoes erupt, magma from deep within the earth is poured out as lava or coughed up as ashes. The intense heat of the eruption resets the $^{40}K/^{40}Ar$ ratio to zero because all of the argon that has accumulated so far is released into the atmosphere. Thus the amount of these isotopes in rocks formed by ancient volcanic eruptions indicates how much time has elapsed since that eruption.

Such dating tells us that *Homo erectus* was alive almost as many as 2 million years ago in Africa. The species disappeared in some regions about 500,000 years ago; in a few localities, it persisted to as recently as about 250,000 years ago. The stone tools commonly associated with *Homo erectus,* at least in Europe and Africa, are grouped under the name the Acheulian industry; the oldest securely dated Acheulian sites are about 1.4 million years old. But while stone tools clearly show that *someone*—some hominid species—was around, they don't tell you who. Hominids are never considerate enough to die with the tools in their hands, and stone tools never die. And if you play the statistical game honestly, identifying the toolmaker as the species most often found with the tools, then the one responsible for making those stone tools is . . . the antelope. The absurdity of this outcome shows you that it is a tricky thing indeed to figure out who made which tools.

Weidenreich's triumph was that he was able to put together a synthesis of the fossil man data that placed the material in an evolutionary framework. His hypothesis, that the regional or racial characteristics of modern humans had begun developing in ancient times, was even testable. As new fossils turned up, he predicted they would display these anatomical patterns. But then he didn't expect his fossils to be lost, which meant future investigations of the Zhoukoudian *Homo erectus* fossils would be inhibited by politics. The threat that politics may pose to science and invaluable specimens is real. Everything from overt revolution to political correctness and museum budget cuts may cause the thoughtless destruction of fossils and other sources of information.

I have few qualms about the physical safety of the Nariokotome boy in the foreseeable future. After Richard Leakey assumed the directorship of the National Museums of Kenya, which houses the boy's remains, he instituted a startling new policy: no fossils were ever to travel out of the museum except in extraordinary circumstances. No longer were specimens blithely shipped off to experts in Europe or the U.S., where they would remain for decades and sometimes be mistakenly accessioned as the property of the local university or museum. No more would the fossils undertake voyages by train or plane or ship, with the attendant risks of loss or damage. The experts would have to come to them. At first Richard's new policy posed no inconvenience to me because I was working in East Africa. After I left to take a job in the States, I, like many other scientists, began an annual pilgrimage to Nairobi to see the original materials.

Spending weeks or months away from home is a serious proposition, and it is harder for colleagues whose spouses are not in the field than for me. Because she is a paleoanthropologist too, Pat understands why I spend two or three months a year in the field collecting fossils and in Nairobi cleaning and analyzing them. It is the nature of my job, and our mutual fascination with human origins was part of what drew us together in the first place.

Despite the personal upheavals to researchers' lives, there have been two major advantages to the policy of not transporting the fossils. First, Richard's strategy (followed by his successor) has successfully minimized the loss, damage, or inadvertent appropriation of specimens that belong to the National Museums of Kenya. Second, the migratory pattern of paleoanthropological research has made me and my colleagues plan our work very carefully. I can't just pop into the lab to take a measurement or verify the placement of some feature, because the university that employs me expects me to be teaching or conducting research in my lab for most of the year. I can only look at "my" specimens once a year, if that, and each trip costs thousands of dollars in grant and personal money. I have learned to think out exactly what information my study requires, not only at this stage but at the next one too. I try to plan for every contingency, and work comes in efficient, concentrated spurts. In some ways, therefore, this policy has helped paleoanthropology immensely.

Richard's precautions extend to the housing of the fossils too. At the museum in Nairobi the exhibits are still displayed in the original colonial-style building made of Nairobi ignimbrite, a sort of orange-brown volcanic rock. It makes a lovely scene. Fuschia bougainvillea cascades down the side of a huge avocado tree near the entrance, where a pair of raucous hornbills often sit, testing avocados for ripeness and sending the rejects crashing to the ground like falling bowling balls. Schoolchildren in bright uniforms parade in endless crocodile formations in and out of the exhibit areas. They stop to buy ice cream from vendors out front or cross the driveway to buy cold sodas and squeal at the inhabitants of the snake park before they file back onto their school buses. All the fossil and archaeological materials now reside in another large, white complex tucked away behind the exhibits building. A bronzed statue of Richard's father, Louis Leakey, in front of the research building, portrays him sitting and contemplating a hand ax.

The structure is built around a central courtyard, which is graced by a large fig tree and a profusion of flowering impatiens and tradescantia. Windows overlook the courtyard from all three floors, making it easy to watch the birds and occasional bats that fly in and out to feed. These amenities soften the hard reality: the building is like a fortress to protect the fossils. To reach the hominids you walk past the statue down a ramp to a small patio and then up to an entrance closed with iron grill-work, through which the courtyard is visible. A bored *askari* (guard) stands there, and a smartly dressed receptionist sits in a booth and works the telephones. She asks whom you wish to see or, if you are an old hand like me and many other researchers, she simply smiles and hands you the key to the office or lab where you are working.

My usual office is halfway around the courtyard, through another grill, next door to that of Emma Mbua, the curator of hominids. The only way to the hominid fossils is through Emma's office, literally. Behind it is a small antechamber, which we sometimes use for photography, and then a massive vault door. Only Emma and the director hold the key. As she swings the door open, the hushed quiet of the room we call the "chapel" becomes apparent. Its pure white walls are thick enough to withstand a mortar shell. They rise to a ceiling high above your head that is set with recessed lights. An air-conditioning system keeps the heat and humidity constant. (Like most buildings in Nairobi, the rest of the museum has neither air-conditioning nor heat. In fact, the design is so open and breezy that heat isn't the problem; lack of it is. In the rainy season, the entire building seems unpleasantly cold and damp, as if the cement floors and walls have never dried.) The floor of the chapel is covered in dense carpeting with thick padding beneath, to cushion but not swallow any material that might be dropped. There is a long padded table with goose-neck lamps and often a microscope or two in the center of the room. Around the walls are large gray metal cabinets with double doors, all locked. Emma has the keys.

Inside are the fossils, organized by site and locality. They are contained in pale wooden boxes about the size of a modest suitcase, each labeled to show the specimens within. Each box, closed by a clasp, lies on its side on its designated shelf. When the clasp is undone, the lid lifts to reveal a lining of blue foam rubber, with cutout shapes that nestle each fragment and protect it from damage. Every fragment is numbered, with a code that combines an abbreviation of the site name, such as WT for West Turkana, and an accession number. A card, or a typed list, of the contents of each box gives further information, such as "*Homo erectus* mandible with left M_{1-3}, heavily worn" for a lower jaw of that species with all its teeth missing except for three cheek teeth, or molars, on the left side, or "*Homo erectus,* left femoral shaft" for the midsection of the left thighbone of that species. Emma's permission is required to take a hominid fossil out of the chapel, even the short distance to my office, and, as a rule, each of them has to be returned to its cradle at the end of the workday. Emma guards her charges as carefully as if they were her children.

At night the iron grillwork gates are locked and two *askaris* patrol with *pangas* (machetes) in hand, checking on staff working late, looking for windows left open by mistake, and listening for attempted break-ins. It is more likely that the computers, cameras, and microscopes rather than the fossils would attract thieves. By preference, the *askaris* are usually Turkana, the fierce tribe that lives by the lake in northern Kenya where Richard, Kamoya, and I have found so many hominid fossils. They have no extended families living close by, no relatives of dubious honesty to pressure them into carelessly leaving a door unlocked, as has been the case in a number of domestic break-ins in Nairobi. The *askaris* speak little Swahili and less English, but they have no doubts about their duties. I am always careful to see them and speak to them if I have to come back and work after dinner or on the weekends. I don't want to be taken for a burglar.

Back in the sixties, when the Department of Paleontology was housed in a rabbit warren of long, low, dingy buildings down the hill from the new building, things were rather more informal. There is a lovely story of an English researcher stopping by in the evening to check on a specimen, and being challenged by an *askari*. The *askari* was probably an ex-military chap, for he cried out in deep and menacing tones, "Who goes there?"

The researcher replied airily, "It's all right, my man, I'm *British*," and strode along on his way.

Things are no longer like that in Kenya. The colonial days are only a memory here, as in Beijing and many other parts of the world where *Homo erectus* fossils are found. And during those distant days no one anticipated that Africa, the lost continent as far as paleoanthropological studies were concerned, would be the ultimate home of the missing link. During the 1930s and 1940s, when *Sinanthropus* fossils were being excavated from Zhoukoudian, it became "common knowledge" that Asia was the birthplace of humanity.

It took Dubois and Black to draw attention to Asia, and it took two more men, who never followed anyone else's path—Raymond Dart, an Australian anatomist, and Louis Leakey—to start the search for the missing link in Africa. Neither was a conventional man and neither thought or worked in the usual way; together, they shattered the stereotype that Asia was the place to look for early hominids.

Dart didn't set out to find the missing link, exactly, but like Davidson Black, he had caught the fever from Grafton Elliot Smith in London. Studying with Elliot Smith, a fellow Australian and the famous and internationally respected expert on the brain, had been a dream of Dart's since he was an undergraduate. Dart earned his medical degrees in Australia and then enlisted as a fledgling physician in World War I. Just after he was demobilized, he heard that Elliot Smith needed anatomists at University College, London. Dart was thrilled to get the job, but perhaps because he was an emotional and unorthodox man he never quite fit in in London. Elliot Smith was suave, cosmopolitan, and polished, part of the British academic establishment despite his antipodean origins. He must have found Dart clever, even brilliant, but perhaps a bit of a liability. At any rate, he sent Dart off to the hinterlands repeatedly. First it was to the United States for a fellowship year of teaching, study, and research. Upon his return, Dart dived into anthropological studies with unfettered enthusiasm. Soon Elliot Smith found another far-distant occupation for Dart: the newly established chair of anatomy at the University of Witwatersrand in Johannesburg.

It felt rather like exile to Dart, according to his autobiography, and perhaps it was. He was not welcomed wholeheartedly, for many South Africans were prejudiced against Australians, who were perceived to have behaved brutally during the Boer War, when many Boers were incarcerated in concentration camps under appalling conditions. Dart found the physical plant at Wits, as the university is called, depressing and shabby. The amenities for running a gross-anatomy lab—trifles like water, electricity, gas, or compressed air for experiments—were nonexistent. There were no adequate wrappings for the cadavers and no anatomy books either. Dart's American wife, once a medical student herself, burst into tears on their first inspection of the facility. Dart described his dismay at seeing the pockmarked walls of the vast dissection hall, attributing the damage to students practicing tennis and football.

Gerrit Schepers, a student of Dart's who was later on the faculty at Wits, remembers a different and more shocking set of circumstances. He recalled recently:

> The incidents about tennis ball marks in the anatomy hall did not occur prior to Dart's arrival. They were a consequence thereof. He had inherited a spic and span modern department from Dr. Stibbe . . . his predecessor. . . .

The tennis ball marks were not [from] tennis balls. There were other round objects obtainable from corpses. The bats were femurs. The game was baseball. I caught two of the students in the act and threw them out of the class. Dart put them back in when their parents objected to discipline.

By his second year, Dart had managed to improve conditions and inspire his students with his unusual teaching style. He was a riveting, dramatic lecturer, who might run from tears to podium-pounding in a single hour. Accounts of his swinging by his arms from the pipes in the dissection hall or demonstrating the "crocodile walk" (a reenactment of the evolutionary transition from reptilian locomotion to mammalian) still circulate. "One learnt absolutely no anatomy from him," says an amused Phillip Tobias, Dart's former student and successor at Wits. "One learnt about Dart."

Because there was no reference collection of skeletons, Dart announced a competition among his students to see who could bring in the most interesting and unusual bones, with small cash prizes. In 1924, the only woman in the class, Josephine Salmons, told Dart she had spotted what she thought was a fossil baboon skull on a friend's mantelpiece. Dart pointed out that her assessment was improbable, since only two fossil primates were known from sub-Saharan Africa, but she borrowed it and brought it in anyway. The baboon skull turned out to be just that, and a most important specimen too. The friend was the mine manager of the Northern Lime Company at Taung, where stones and fossils turned up with some frequency, and Dart soon arranged to have any fossils found there boxed up and sent to him.

It is a classic story of anthropology, all the more engaging for being true. With unerring timing, the box with the missing link in it turned up as Dart was dressing, in wing collar and morning dress, to serve as best man and host for a friend's wedding. The men from South African Railways who staggered up to the house that summer day in 1924 left two large crates blocking the stoop shortly before the guests were to arrive. Dart had them moved to the pergola, where they would be out of the way, and left off dressing to find a crowbar to pry them open. The contents of the first box were uninteresting scraps of fossil eggshells and turtle scutes (the bony plates that underlie the turtle's shell). On the top of the rubble that filled the second crate, Dart spied an extraordinary thing: a natural, fossilized cast of the brain—an endocast. It was of some creature whose brain was about as big as that of an adult chimpanzee. From his work with Elliot Smith, Dart immediately recognized that this was no ape endocast (unparalleled as that would have been), but one with distinctly human anatomy. He rummaged through the box frantically and found a piece of bone, covered in rock, into which the endocast fit. And then real life intervened. The groom appeared, anxious that Dart should brush the dust off his suit and struggle into his stiff collar; the wedding party was arriving momentarily. Dart took these two precious pieces of our ancestry and locked them in his wardrobe, reluctantly abandoning them until the festivities were over.

There was nothing else in those crates to rival the Taung child, as it was soon nicknamed. Dart took months to clean the fossil, using a small hammer and chisels and his wife's sharpened knitting needles. When he finished he had one of the most beautiful gems of fossil hominids, an exquisite thing: the intact face, with lower jaw and endocast, of another missing link—not *Homo erectus*, but an earlier, ape-ier creature that Dart named *Australopithecus africanus,* the southern ape-man of Africa. Although the bony braincase itself was broken away, its shape and size were clearly delineated by the endocast. The teeth were all in place and showed that this individual was a youngster at the time of death; today anthropologists age the Taung child between three and a half and five years old in human terms. By early 1925 Dart was ready to introduce his child to the world. He chose the time-honored route of publishing a paper in the premier British scientific journal, *Nature*. Never one to shun bold moves, Dart proposed that an entirely new zoological family, the Homo-simiadae (the man-apes), be erected to receive this specimen.

Initial responses were favorable. Elliot Smith sent a cable congratulating Dart, as did Aleš Hrdlička, the "fossil man" man at the Smithsonian, and General J. C. Smuts, who had just completed his first stint as prime minister of South Africa. Newspapers and magazines wanted articles and statements from Dart, but the scientific establishment in London was miffed at being taken by surprise. Despite its relatively vertical face and small teeth, two of the attributes that distinguish humans from modern apes, these critics wrote of the terribly *ape*-like quality of the skull. It was, they observed, a youngster, a child, and the young of apes are notoriously more humanlike than the adults; another way to look at this is that humans have prolonged the duration of the childhood traits of apes into adulthood.

A mere week after the initial praise, articles appeared in *Nature* criticizing Dart's ideas and flatly contradicting his conclusion that *Australopithecus africanus* was a missing link between apes and humans. All were written by authorities who had seen neither the original nor a cast of it. Their sole source of information was Dart's own *Nature* paper and the photographs that appeared in it. Leading the pack was Arthur Keith, a Scottish anatomist at the Museum of the Royal College of Surgeons, another eminent scientist who was deeply involved in the Piltdown debate.

The issue was that the faked Piltdown skull "proved" Keith's pet theory that the fundamental adaptations of humans were their large brains. He was so convinced of this theory that he even proposed an explicit cerebral credential for inclusion in the genus *Homo*: a cranial capacity of 800 cc or more. Because Keith was certain that the large brain had evolved very early in human evolution, while the teeth were still large and primitive and the face long and protruding, he believed the evidence of the Piltdown skull without hesitation. The forger, whoever he might have been, pandered to Keith's convictions. He used pieces from a few-thousand-year-old human cranium combined with an artfully broken and stained orangutan jaw, with its large teeth cleverly filed to resemble a more human pattern of wear.

These specimens had been known and accepted by many for more than ten years when Dart's Taung child turned up to show the opposite condition: a small, ape-size brain, yet one with some genuinely human attributes, coupled with extremely human teeth and jaws. The Piltdown and Taung skulls could not both be direct human ancestors, for either the brain had evolved first or else the teeth had. Keith, and others who agreed with his theory, rejected Dart's find. After all, they all knew Dart was an emotional, even unstable type, given to hyperbole. Keith even wrote in his autobiography:

> I was one of those who recommended [Raymond Dart] to the post [at Johannesburg], but I did so, I am now free to confess, with a certain degree of trepidation. Of his knowledge, his power of intellect and of imagination there could be no question; what rather frightened me was his flightiness, his scorn for accepted opinion, the unorthodoxy of his outlook.

Dart had probably mistaken a juvenile ape fossil for a human ancestor, Keith thought, and said so. The rejection of Dart's discovery was very similar to Dubois's bitter experience.

About the only believer Dart attracted was another physician-cum-paleontologist, an eccentric Scot named Robert Broom, who in 1925 was working as a country doctor in a series of small South African outposts. Broom also had a brilliant and well-respected career as a paleontologist—he was given a medal by the Royal Society in London for his research—and cared nothing for prevailing paradigms. The story is told that he burst into Dart's lab one day and, without bothering to introduce or explain himself, rushed across the room to fall on his knees before the Taung child, in rapt contemplation of our ancestor. Even though no one else believed Dart was right, Broom did.

Winning Elliot Smith's approval and endorsement was clearly on Dart's mind as he took his fossil to London in 1930. But the Taung child was upstaged by Davidson Black, visiting from China with his new Peking Man fossils and a professional slide presentation for scientific meetings. Dart's precious find was not only overshadowed, it was literally abandoned—left, in its humble brown-paper-covered box, in the backseat of a London taxi by Dart's wife. It was recovered only after frantic searching. Dart gave up on plans to publish a monograph and returned home, discouraged and defeated. He gave up fossil work for many years and subsequently suffered a nervous breakdown. For years, the Taung child sat, forgotten, on Dart's colleague Gerrit Schepers's desk at Wits, in the medical school surgery unit where he was doing experiments on baboon brains.

Broom suffered no such loss of confidence, even when his convictions checkered his career. Some fifteen years before meeting Dart, Broom's insistence on evolution had cost him his professorship in geology at the University of Stellenbosch, an extremely conservative institution with strong religious ties. As one memoir describes him, "It seems likely that then [1910] as in later years he would never have stepped an inch out of his way to avoid a fight, but rather would have taken a perverse delight in steering a collision course. . . ." Broom had been unable to secure another academic post, as all the South African universities were part of the government system, so he started practicing medicine in the lonely Karroo region, which incidentally full of the dinosaurian and reptilian fossils that most fascinated him. He continued his paleontological studies at a furious pace and wandered as an itinerant doctor from one small town to the next.

In 1934, at the age of sixty-eight, Broom "retired" from his medical career and took a post at the Transvaal Museum in Pretoria. Not long afterward, Schepers, who was on the faculty at Witwatersrand, and one of his masters-of-science students, Harding le Riche, approached Broom at the museum with a fragment of mandible that they thought belonged to an early human ancestor; they had recovered it from a cave at Sterkfontein. Broom sent his visitors away, saying they were wasting his time, but they returned again some months later with another jaw fragment, this time with some teeth in it, and caught his attention. Broom asked them to take him to Sterkfontein on the next free weekend, which proved to be August 9, 1936. Broom, as ever, was clad in a dark three-piece suit and wing collar. "This was not inappropriate dress at all," Schepers asserted recently, recalling the occasion. "Broom had been a medical man all his life, and it was important to preserve a certain formality of dress. He used to buy very expensive suits, which lasted a long time because they were made of good strong wool. The new ones he saved for the office and the old ones he used for his expeditions."

Broom was enormously impressed with Sterkfontein's potential, especially when Schepers showed him a snout—upper and lower jaws—that he had partially cleaned from the rock. Although the specimen is not in the Transvaal Museum, Schepers believes this was the very first australopithecine recovered from in situ at Sterkfontein. Broom, typically impatient and full of energy, returned the next day with a full museum crew and started excavation at once. Schepers and Harding le Riche, who had started the paleontological work at Sterkfontein, were brushed aside. Within a few days, the manager of the quarry at Sterkfontein, G. W. Barlow—who had also been in charge at

Taung—handed Broom a fossilized braincast of an adult australopithecine, the first of many hominid fossils from Sterkfontein. "Is this what you're after?" he asked. It was.

Broom knew the importance of brain size in assessing the humanness of the remains. He worked with a young assistant named John Robinson, a zoologist who supervised the crew and prepared and helped analyze the fossils. Over the next few years, in caves like Swartkrans and Sterkfontein, Broom and Robinson found *adults* of creatures like the Taung child. They were not just apes. The most famous specimen was nicknamed Mrs. Ples, a contraction of the formal name Broom gave it, which was *Plesianthropus transvaalensis* (near-man of the Transvaal). When the time came to write up these fossils, Broom asked Schepers, who was by then an expert on the human brain, to describe and analyze the braincasts of the Sterkfontein australopithecines, while he and Robinson dealt with the bony remains. Although he multiplied names as fast as anyone in this era, Broom actually found two species: a gracile, or lightly built, form that is classified today as *Australopithecus africanus* (like the Taung specimen or Mrs. Ples) and a robust, crested-skull creature with huge teeth and jaws that is now called either *Paranthropus robustus* or *Australopithecus robustus*.

Like many scientific endeavors, Broom's fossil hunting was interrupted by World War II. But in 1946, General Smuts, once again the prime minister, called upon Broom to resume the search for the missing link, with government funding guaranteed. Broom and Robinson had no sooner started in again than Broom received word from the Historical Monuments Commission that he would not be permitted to carry on his work without the assistance of a "Competent Field Geologist." He was perceived as being far too focused on getting the fossils out—by blasting, pickaxes, or whatever crude method worked—and seemed to be taking no notice of the stratigraphic context. Broom, a one-time medalist in geology at Glasgow University and, for seven years, professor of geology at Stellenbosch, was incensed; he had no intention of complying.

> Gen. Smuts was in America [recounted Broom] and too busy to be appealed to. I was quite ready to defy the Historical Monuments Commission, and carry on. It would only have meant at the worst a fine of £25. I felt sure the Commission would not dare to take the matter to court. In fact, I was very anxious that they would; and had engaged a barrister to defend me. I had no compunction whatever about breaking the law. I considered that a bad law ought to be deliberately broken.

Smuts returned and undertook to smooth things over. When Broom received a permit to work at a site at Kromdraai, but not at Sterkfontein, he deliberately moved his operations back to Sterkfontein to flaunt the Historical Monuments Commission regulations. Broom and Robinson soon found good specimens, but the attendant publicity led to Broom's being strongly condemned by the commission. He blithely restarted work at Kromdraai. A geology professor from Pretoria University was called in to evaluate the "damage" Broom had been causing; in Broom's words, the geologist declared "that there was no stratigraphy whatever at the place where I had been working, and that I had been doing no harm." With this, the commission tacitly accepted that there was no way to control Broom and gave up.

Of course, there *was* stratigraphy in the caves, and Broom *was* absolutely ignoring it. The situation was extremely complex, as later painstaking studies by C. K. Brain would show. The caves had been formed as underground solution caverns. Bones, sediments, and other debris had fallen into each cave through holes in the roof, with periodic roof falls that altered the size and location of the holes considerably. Under each hole, the sediment and bones didn't form flat, layer-cake strata but made a cone-shaped hump, so that new material might land on the cone and roll down the side (where it would lie lower than the older material) or it might hit the cone and settle in. None of this was obvious to Broom or anyone else at the time.

In another cave known as Swartkrans, Broom and Robinson found some odd specimens that didn't look like the rest of their hominids, who were in themselves a pretty diverse bunch. Their suspicions that yet a third species was present started in 1949 with a mandible found by Robinson, which was more gracile and had much smaller teeth than those of the robusts. It seemed far more human than the other jaws. Soon there were some distinctive fragments of face and upper jaw as well as the end of a radius, one of the bones from the forearm. The face seemed more vertically oriented than robust faces, and the radius was very human indeed. They called it *Telanthropus capensis,* the distant or far man from the Cape; it was actually *Homo erectus*, but they didn't know it. Part of the reason was that Swartkrans, the cave *Telanthropus* came from, was full of robust australopithecines; no one expected an early human to show up with the most primitive hominid we knew. And, although a large part of one individual's skull was there in pieces, no one realized that the facial fragment fit onto the braincase bits (some were still classified as australopithecines) until a British anthropologist, Ron Clarke, reworked the material in 1970.

Because of Broom's work, and the belated but clear support offered by the famous British anatomist Sir Wilfrid E. Le Gros Clark—one of the few who actually went to South Africa to look at Dart's precious find—in the early 1950s the australopithecines were finally acknowledged as hominids. It must have been a long twenty-five years from Dart's perspective. But at the end of this trying period, Dart had the satisfaction of knowing that he had established the original homeland of the missing link as Africa, not Asia. His Taung skull preceded Black's

Peking Man by hundreds of thousands of years and cast serious doubt on the primacy of the big brain in human evolution. This left open the question of the nature of the fundamental and original hominid adaptation.

Other australopithecine fossils found by Broom and Robinson, and analyzed by the latter after Broom's death, offered an alternative candidate for that adaptation, for these fossils showed clear anatomical signs of having been bipedal. Somehow, bipedalism wasn't nearly as satisfying a criterion for election to the elite human club as braininess, especially when membership was voted upon by a species with the audacity to dub itself *Homo sapiens*, or "man the wise." This implicit preference for brains over legs probably made it all the harder for Dart's precious australopithecines to win acceptance.

So for many years, Dart's ape-man was completely denied as a hominid. In Louis Leakey's autobiography of his early years, *White African,* written in 1937, he describes the events of 1925, the year Dart first wrote about *Australopithecus.* At the time, Louis was busily trying to persuade the world that Africa was a good place to seek early human ancestors, yet it is telling that he mentions neither Dart nor his new find from South Africa. Louis was running some of his early expeditions—wild and woolly affairs, they were too—in remote and untamed regions of the colonies that became Kenya and Tanzania, where he found prehistoric sites and even burials, but they were of anatomically modern humans. To him, the Taung specimen said nothing about the roots of humankind. Indeed, his search for the missing link in Kenya owed nothing to Dart's discovery.

If you want to understand Louis Leakey, you have to realize that he was the son of a pair of missionaries in colonial Kenya. His formative years were largely spent in Kenya, making Leakey, like Dart, a bit of an oddball in England. Louis trained in anthropology at St. John's College, at Cambridge—my own college years later, which made it easier for him to "forgive" my excavating fossil sites in Uganda that he might have been interested in himself. Louis was considered a most promising young scholar and worked for a while under the tutelage of Arthur Keith, the London anatomist who had been so deeply skeptical of Dart's Taung child. Louis sought the older man's help in describing some of his skeletal finds and simply absorbed Keith's "brain first" idea of human evolution.

Louis had several immense advantages for working in East Africa: he spoke Kikuyu (he was initiated into the tribe at age thirteen, with his age-mates) and Kiswahili fluently, as well as bits of other local languages; he knew the country and how to get around in it; and he was a part of the small but scattered community of white settlers, who would send word when they noticed fossils or artifacts or offer accommodation at their farms when he was working nearby. He knew that early hominids had lived in East Africa because, as a boy fascinated with prehistory, he had found their tools—even if the academic community in England didn't know it. Louis was also a tremendous showman, who knew how to work the English academic networks. At this stage in his career, he was very successful in obtaining fellowships, grants, and other opportunities despite his radically un-English upbringing. Unlike Dart, who seemed an outsider, the English establishment treated Louis as eccentric—a colonial—but one of their own.

Louis began fossil hunting in Kenya during the 1920s when his doctor recommended taking a year off from Cambridge to recover from a concussion. Probably a rough-and-ready expedition to Africa was not the rest cure the doctor prescribed, but Louis, like every other Leakey I have ever met, had unshakable confidence in his own opinion. He sought out a German, Hans Reck, who had prospected an immense gorge in Tanzania (then Tanganyika) and reported finding fossils. It was known as Olduvai Gorge, after the abundant *Sansevieria* plants called *ol tupai* by the local Masai.

The gorge has a series of layer-cake beds, numbered I (at the bottom) to IV (at the top). Their relative ages are obvious—Bed I was deposited first—but their real ages were entirely unknown because there was no means of absolute dating at the time; rough guesses were usually made on the basis of associated faunas. In 1913 Reck had found and excavated Oldoway man, a complete skeleton in the knees-drawn-up position that anthropologists call a flexed or crouch burial. Although some crouch burials are as old as perhaps 100,000 years, they are usually typical of more recent, modern human burials, like those Leakey had found at Gamble's Cave, in Kenya, which he guessed were only about 20,000 years old. So how did Oldoway man, an anatomically modern human, come to be interred in Bed II? The associated fossils were of Pleistocene animals, suggesting an age of perhaps 750,000 years old. Was this an intrusive burial, dug into much older layers?

In 1931, Louis revisited the site with Reck and some geologist colleagues to assess the situation. At the time, they adjudged the sediments above the skeleton to be undisturbed, meaning it was *not* intrusive; later, when Louis became suspicious again, fluorine studies proved the burial was both recent and intrusive. The fossilized animals were as ancient and abundant as promised and, to Louis's great glee, he was able to win a long-standing bet with Reck: he found a primitive stone tool within twenty-four hours of arriving at Olduvai. Why had Reck failed to recognize these tools? As a European geologist, Reck was familiar with flint tools, which are often finely worked; the crude stone tools flaked out of lavas and quartzites that are found at Olduvai simply looked like lumps of rock to him (and, I admit, sometimes to me too; I've been accused by archaeologists of leaving my footprints next to stone tools when I'm looking for fossils).

Most paleontologists have one or maybe two sites that make their careers—that is, if they are lucky—and Olduvai was Louis's. It was the source of his well-deserved popular and scientific acclaim. Those who had never heard of him until 1960, when he and his extraordinary wife Mary published the first of many *National Geographic* articles, didn't realize the magnitude of the accomplishment. From 1931 until 1959, Louis and various colleagues worked at Olduvai, four grueling days' drive (when the trip went well) over nonexistent roads from Nairobi. They drank rhino-urine–flavored water, had nasty encounters with lions, ran short of food, and sometimes sank so low as to scurry around collecting their discarded cigarette butts when they had no more cigarettes. They made roads, made camps, and made do. The field life spurred the demise of Louis's first marriage and was in large measure the foundation of his second, to Mary Nichols, an unconventional, sharply intelligent woman who was working as an archaeological illustrator at the time. Louis's academic prospects and ability to obtain grants were seriously damaged by his marital upheavals, since divorce was a terrible scandal in England in the 1930s. Even worse, he and Mary had lived together in England without benefit of clergy for a year while awaiting his final decree. After that, for many years, there were only tiny grants or no grants; Louis's expectations of an academic position at Cambridge disappeared. Louis's marriage to Mary was nonetheless a good partnership, with Louis's charm and flair for public speaking balanced by Mary's quiet persistence and scientific acumen. He had the ideas; she collected and documented the evidence.

It was Mary who found the fossil that changed their lives and professional fortunes. She spotted the skull one morning while Louis was down with the flu. They dubbed the specimen Dear Boy and numbered it OH (for Olduvai Hominid) 5. It was also known as Zinj for short, from its formal name: *Zinjanthropus* (*Zinj-* an old name for East Africa, and *-anthropus*, meaning man) *boisei* (to honor Charles Boise, a London businessman who helped fund the expedition). Louis always liked to have a new name for his fossils. Journalists called it Nutcracker Man, because of its enormous jaws and teeth. It was like the specimens Broom had found and called *Paranthropus robustus*, only bigger and more exaggerated in anatomy. Today, I would call Louis's find *Australopithecus boisei* and Broom's *Australopithecus robustus*. Together these species are sometimes loosely called the robust australopithecines, but the East African species is more extreme anatomically, even hyperrobust. Zinj sported a broad and long face; huge, flat molars; and an enormous bony crest that ran front to back on the top of his head for the attachment of massive chewing muscles.

The Leakeys found Zinj after working at Olduvai for almost thirty years. They had found fossils year after year—in some places in Olduvai they are literally thick on the ground—and there were plenty of stone-tool sites to keep their archaeological instincts twitching, but there had been precious few homs and fewer grants. I would have given up years before they did— Mary's patience and perseverance exceeds mine (and Louis's) by an order of magnitude. Yet somehow, even if you are an impatient person like me, when you know there is something good waiting, you come back, again and again, to scrabble in the heat and the dust for that elusive glimpse of your past that reveals so much about your present. There is a strange sort of saneness and simplicity about living in the desert with a handful of people who share your peculiar obsession, as isolated from everyday life as if you had gone to Mars. So, for some of us, it is almost impossible to walk away for good, and you wake up one day and realize you have spent decades of your life on this quest. There are no shortcuts, only determination and luck and trying again. That makes the finding of a good fossil all the sweeter, as it was for Louis and Mary when Zinj turned up. Once Zinj was in hand, the National Geographic Society started supporting their efforts year after year, paying salaries, buying tents and equipment, and purchasing what Gil Grosvenor of the NGS once described to me as "every Land Rover in East Africa, twice over."

Louis and Mary never lived in luxury and never accumulated any personal wealth, but they could get the job done and that was what mattered. Serious excavations began at Olduvai and continued well after Louis's death in 1972. Mary only closed up camp and officially retired in the 1980s, when she suffered from a prolonged spell of ill health (including a fractured ankle) that left her feeling temporarily miserable.

Zinj was just the beginning at Olduvai. This find reinforced the message that the fairly recent acceptance of Dart's Taung child had first promulgated: Africa was the birthplace of humanity and the home of the missing link. Like the Taung child, Zinj showed that a big brain was not the first attribute to evolve in the human lineage (although Louis clung to the brain-first notion and recanted his claim that Zinj was a direct human ancestor as soon as he could find an excuse). Its discovery redirected the quest for the missing link from Asian to African sites, a formidable accomplishment in the face of considerable skepticism.

But the most important thing these scientists did was less tangible if more influential. They sketched the outline of a new view of human evolution, one that started not with massive enlargement of the brain but with changes in the locomotor mechanisms (the pelvis, legs, and feet) and in the face and teeth. It was not thinking so much as moving and eating that mattered in the earliest days of our lineage. Because Zinj and the Taung child were more primitive or more apelike than *Homo erectus,* they changed the context in which *erectus* was viewed. It was no longer a missing link that was wedged between apes and humans, occupying a place on the phylogenetic tree with a living form on each side. Now *erectus* lay between humans and another once-missing link, the australopithecines, a group that was already more humanlike than apes. There were now at least four links in the chain rather than three. This situation lent a much greater time depth to the entire evolutionary process.

Even before Zinj was found, Louis was searching for a better method of dating his material, so he involved Jack Evernden of the University of California at Berkeley. Radiocarbon dating had been invented in the 1940s, but the radioactive decay of carbon 14 to the ordinary carbon 12 happened too quickly to yield a useful date on anything as old as the bones at Olduvai. But in the 1950s Evernden had applied the same principle to another system, tracking the radioactive decay of potassium into argon, which might be useful for dating the geologic beds at Olduvai in which the fossils lay. Louis was initially suspicious but got Evernden to come to Olduvai and collect samples nevertheless. The new technique worked and transformed everyone's ideas about the antiquity of the human lineage. In 1961 Evernden, Louis, and Garniss Curtis announced the date of lower Bed I—and Zinj—to be between 1.6 and 1.9 million years. This first date caused amazement; no one had had any idea these Pleistocene beds were more than a few hundred thousand years old.

Radiometric dating is absolute: you get a number, which you hope reflects to the geologic event in which you are interested. It is the technique that lets you answer the "when" questions, if not the whos, whys, and wheres. Another tremendous advantage to this technique is that you obtain a measure of the accuracy of that magic number too, called a *dating error*. This is a figure, written with a plus and minus sign in front of it, that indicates how precise or sloppy that particular date is. The bed in which the Zinj skull was found is bracketed by two tuffs, or ancient volcanic ash layers, which have recently been dated using today's more sensitive techniques. The lower tuff is 1.80 ± 0.01 million years and the upper one is 1.76 ± 0.02 million years, making Zinj roughly 1.78–1.79 million years old.

More hominids followed, and the Leakeys seemed to name a new species annually. In 1961 Louis found a skullcap more than a little reminiscent of Dubois's. It was Olduvai Hominid 9 (OH 9), the first African *Homo erectus,* but Louis wouldn't accept it.

From his days with Arthur Keith, Louis was sure that he would find an early, big-brained but primitive human at Olduvai, so OH 9 was rather a disappointment. But, in Louis's mind, if brains didn't make our ancestors human then it must have been tools, and he did believe OH 9 was the maker of a primitive stone tool culture, older than the Acheulian industry, which is dominated by teardrop-shaped tools called hand axes. Actually, Louis was wrong about this (the culture was Acheulian), as he was wrong in identifying OH 9 as *Homo* but not *Homo erectus*. The pity is that he never appreciated the best fossil he ever found himself. He steadfastly maintained that *Homo erectus* could not be a direct ancestor of modern humans because its brain was too small too late in time. It was nothing but an extinct side-branch to Louis. He missed the missing link.

But he found other bones, including some from a fossil hand that reached out and touched my life. Between 1961 and 1963, the Leakeys unearthed some hand bones, a jaw, and some skull fragments that Louis thought belonged to a single juvenile individual, OH 7. Louis didn't quite know what to do with this collection, so he called for expert help. The skull was sent to Dart's successor at Wits, Phillip Tobias (also known as PVT, from his initials); the postcranial elements went to John Napier at the Royal Free Hospital, where I was studying. As John always did, he shared this new material with all his graduate students, including me. The excitement of their analysis—the process of squeezing the maximum information out of a set of bones—made a deep impression on my life. It was probably then that the missing link took hold of me, even if I didn't know it yet.

6

SEX AND THE MISSING LINK

I hadn't thought much about it at the time, of course, but even missing links have to come from somewhere; they, too, had to have ancestors. It wasn't long before the new find from Olduvai, OH 7, was nominated for this position: the species from which *Homo erectus* had evolved. When Louis Leakey involved my adviser, John Napier, to help him analyze OH 7, I found myself in the enviable position of watching them discover what these new finds meant.

As Louis, John, and Phillip Tobias wrote up the fossils for publication, Louis's dominant personality and rampant enthusiasm may well have influenced the others to downplay the anatomical evidence, emphasizing instead the implications of their observations for the humanlike attributes of this new specimen. Wanting to hammer home the message that this new material represented a species that was the toolmaker at Olduvai, they sought a new name. Although he had yet to see the specimen, Raymond Dart happily suggested *Homo habilis,* meaning "handy man." The three scientists' description of the new species represented by these fossils, full of ideas about behavior and humanness, was published in 1964.

John had found that the Olduvai bones were remarkably modern in their anatomy, fully capable of two different, and fundamentally important, types of grip. He coined two terms, *power grip* and *precision grip,* to indicate different manual capabilities that he had first come to understand when he was working as a hand surgeon during World War II. As he strove to restore function to soldiers with hand wounds, he discovered which particular anatomical structures made which sort of action possible. Ever afterward, his strong point was drawing the connections between form and function in limb bones. In John's lexicon, which quickly became the standard, a power grip is what you use on a hammer or a screwdriver, or any object that you clasp to your palm with your fingers and can exert some force upon. All primates and many other mammals use this sort of grip. In contrast, a precision grip is the daintier, fingertip hold you use on a pencil or a fine implement the movement of which you want to control precisely. The precision grip requires an opposable thumb, a thumb with an unusually shaped joint at its base that allows it to be rotated so that (if the thumb is long enough) its fleshy tip meets the ball of another finger. As John pointed out, the precision grip and the long, opposable thumb are very human acquisitions, important prerequisites to effective toolmaking. (Though chimpanzees and other apes manage to make and use very simple tools, their thumbs are opposable but so short that objects are held between the thumb and the side of the index finger, making their movements awkward and limiting their fine control over the tools.)

John's analysis of the anatomy of the new finds was that the species from which they were drawn—like modern humans—

had both power and precision grips, a perfect combination for the earliest toolmaker. This finding delighted Louis no end, since he had never been comfortable with the hypothesis that the small-brained, heavy-skulled Zinj was the toolmaker at Olduvai.

When a new species is named, a type specimen is designated that carries the species name; it is the embodiment of the technical diagnosis of the specimen. The type specimen of *Homo habilis*—the fossils to which they formally attached the new name—included not only the hand bones John worked on but also a mandible, an isolated tooth, and some skull bones, all of which were believed to come from one individual: OH 7. All of these bones were from an immature individual, but, apart from that, the evidence linking them to one original skeleton was thin. They were not found particularly close to one another nor—because some pieces were from the skull while others were from the hand—could they be rearticulated. Other hominid specimens that were listed in the original paper as belonging to *Homo habilis* came from several other individuals and localities that were used as demonstrations of other parts of the anatomy of this new creature. From another individual, a young adult, there was a complete mandible and a palate, upper teeth, and parts of a braincase; these fossils were numbered OH 13 and known as Cinderella or Cindy. Additional hand bones, including two finger bones that were later realized to have come from a giant colobus monkey; some foot bones of an adult; some odd teeth; and a few jaw fragments and bits of skull rounded out the evidence of *Homo habilis*. Judging from this motley hodgepodge, the new species had relatively small teeth (much smaller than Zinj's, in any case) set in a dental arcade, or tooth row, that was approximately parabolic, rather than describing a square or rectangular shape as in Zinj. Both these features argued for a more humanlike condition. More evidence came from PVT's inspection of the skull fragments from the type specimen. Although OH 7 was too incomplete to make a convincing estimate of its cranial capacity, PVT felt that the braincase was larger and more globular than Zinj's. And if *habilis* had a larger cranial capacity than Zinj, then logic suggested that it must also have had more nearly human capacities than did Zinj. These conclusions, like John's, reinforced Louis's preconceived notions.

Taxonomy, the science of naming things, is a formal and stuffy business. It works by extremely precise and conservative rules, one of which is that any new taxon, or name, must be presented with an exact anatomical diagnosis that lists the defining characteristics of that taxon. Therefore, the boldest and most unorthodox move of Louis, John, and Phillip was revising the diagnosis of the genus *Homo* to shoehorn this new specimen in, based primarily on questions of brain size. For a long time, following Keith, the diagnosis of *Homo* had included having a braincase bigger than 750 cc, a point that amounted to a specific definition of what "big-brained" meant for a human- or ape-size species. To Keith, and to Louis after him, humanness lay largely in brain size, which both agreed must have evolved very early in our ancestry. Louis's new cranium could not be inflated to encompass 750 cc no matter how PVT reconstructed it; in fact, PVT's rather generous, preliminary estimates of the cranial capacity of the type specimen ranged from 643 to 723 cc for *Homo habilis*. (While this represented a moderate increase in brain size over Zinj's 530 cc, the gap between them has since been eroded by more recent finds of other skulls with intermediate-size brains.) The only solution was to enlarge the range of brain sizes in the genus *Homo* so that it would easily encompass this new specimen, and so Louis, John, and Phillip did. The revised diagnosis of the genus *Homo,* as given in their paper, included the following remarks:

The cranial capacity is very variable but is, on average, larger than the range of capacities of members of the genus *Aus-*

tralopithecus, although the lower part of the range of capacities in the genus *Homo* overlaps with the upper part of the range in *Australopithecus;* the capacity is (on average) larger relative to body-size and ranges from about 600 c.c. in earlier forms to more than 1,600 c.c. . . .

Because of John's analysis, the diagnosis also included the possession of an opposable thumb and the ability to use both power and precision grips. Thus they were discussing behavioral criteria rather than strictly anatomical ones for inclusion in the genus *Homo,* a radical step that blurred the distinction between the facts and the interpretation and made the diagnosis a difficult one to use later.

Although the evidence of cranial capacity was weak—how strong a credential could cranial capacity be if it could be altered with each new specimen?—it was propped up by the anatomical evidence of technical adroitness from the hand bones, which in turn seemed to be supported by the association of the fossils with abundant stone tools, especially OH 13 (Cindy). As I said earlier, stone tools are tricky things; they never die and they never go away, so they end up associated with all sorts of fossils—and these particular stone tools were found with both Zinj and the new *Homo habilis* fossils. There was no special reason to decide that one and not the other, or both, was the toolmaker except that the *habilis* skull seemed to have a larger brain and more humanlike teeth; the paper (sounding a great deal like Louis) simply asserted that "it is probable that the latter [*Homo habilis*] was the more advanced tool maker and that the *Zinjanthropus* skull represents an intruder (or a victim) on a *Homo habilis* living site." No evidence supporting this probability was offered. Louis simply believed *habilis* better to be the toolmaker because it was more "advanced," meaning "bigger brained" or more like us.

Thus, in this paper, Louis, John, and PVT shifted the ground rules twice. Not only did they redefine *Homo* so that the new fossils could be accommodated in that genus, they also suggested *behavioral* criteria rather than the traditional anatomical ones to support their diagnosis. They overturned previous opinions (that *Homo* had a brain bigger than 750 cc; that the toolmaker at Olduvai was Zinj) with an ease that outraged the scientific community. What suddenly made 670 cc adequate for the conceptual tasks involved in toolmaking when the consensus had always been that 750 cc or even 800 cc were necessary? Louis, John, and Phillip were criticized by nearly all of their peers, a reaction that Phillip attributed to the fact that the *habilis* paper shifted the mental paradigm under which paleoanthropologists had been operating. It was rather akin to getting halfway through a game of Monopoly with someone who then changes the rules. But the paper was also poorly conceived, and because it stands forever as the formal diagnosis of *Homo habilis,* the species has represented a problem for paleoanthropologists ever since.

Emotional convictions like the ones that motivated Louis with regard to *Homo habilis* are hard to avoid. Fossils that you have found, particularly skulls, are like your children. That is why fossil finders guard them so jealousy and love to give them new names. Perhaps the most transparent manifestation of this view was an action taken by Eugène Dubois, who had his own eleven-year-old son pose naked, standing unhappily in the chilly attic of an old building while Dubois sculpted a statue of *Pithecanthropus erectus* to be exhibited at l'Exposition Universelle in Paris in 1900. Nothing could have made clearer the symbolic identity between offspring and fossil—and Louis felt just the same. So, since he prized his own cleverness above all else, surely Louis's fossil would be an "intelligent" one.

In an odd way, the most diagnostic characteristic of *Homo habilis* is that it fills the evolutionary gap between australopithecines and *Homo erectus* (although Louis would turn in his grave at the thought that *Australopithecus–Homo habilis–Homo erectus* represented a single evolutionary lineage, because he thought *Australopithecus* was too small-brained to be ancestral to *Homo*). In other words, this new species was defined more by its neighbors than by its own distinctiveness. Since *Homo erectus* has a brain roughly twice as large as (or larger than) that of australopithecines, it is inevitable to imagine the existence of a species with a brain of intermediate size. Such expectations are seductive and sometimes dangerous. Haeckel constructed the idea of *Pithecanthropus* out of no more than his expectations of the transitional form between humans and apes—leaving the reality to be discovered by Eugène Dubois in Java. To a lesser extent, Louis, John, and PVT constructed *Homo habilis* as a link between australopithecines and *Homo erectus,* mixing up a potent cocktail of their expectations enlivened by their anatomical observations. Even though they had a mandible, some parts of the skull, a foot, and some finger bones, they really didn't have enough parts that were distinctive from other known species to name a new species. In my view, it still isn't clear that all of the bits came from a single species anyway, because *habilis* remains so poorly known. Thus, even though more specimens have been found and attributed to *Homo habilis,* the confusion remains so great that no two anthropologists have the same list of specimens that they would classify as *Homo habilis.* The 921-page monograph on the Olduvai *Homo habilis* fossils that PVT published in 1991 did not resolve the uncertainties that first arose with the announcement of the species in 1964.

A year later, as the controversy over the *Homo habilis* paper was in full cry, I set off to take up my first job in East Africa. I was not looking for missing links and I did not expect to work on *Homo habilis;* I was more interested in the ancestors of the lemurs and prosimians I had undertaken to study for my thesis. On my weekends and vacations, I began to excavate a twenty-odd-million-year-old Miocene site in Uganda, looking for such fossils, and promptly fell afoul of Louis. I doubt he realized that I was John's student or that I was a graduate of the same college he had attended at Cambridge, ties that in the British academic network made us relatives of a sort. But of course he didn't like some young upstart Englishman working only one country over on primate fossils that he himself might decide to be interested in. Louis's solution to the problem was to write a high-handed letter to the Ugandan president, Milton Obote, accusing me of incompetence and of damaging important fossil sites. Fortunately for me, the letter went first to the official in charge of granting excavation permits, a friend of mine who conveniently lost the letter. Like most alpha males—the term anthropologists use for the dominant primate in a group—Louis was territorial and did not like poachers on his turf. And, like any nonhuman primate, he could smell an aggressive young male from a long distance.

I liked the fossils, I liked Africa, and I liked my new colleagues too, so I decided simply to stay out of Louis's way. One of my friends at the medical school of Makerere University College was a young Kenyan named Joe Mungai. He was a talented young researcher with a Ph.D. from University College, London, and we got along well teaching together for several years. Then in the late 1960s he received a letter asking him to return and start up a department of anatomy for a new medical school in Nairobi. It would mean a struggle and a heavy administrative load, both of which would pull Joe away from his exciting research, so he declined, explaining his reasons. A telegram followed. It said, simply, "You will come," and he went. Because he asked me to join him, I went too. Joe was right about his future: he became vice-chancellor of the University of Nairobi and is now an administrator in higher education in Kenya. He never returned to his research.

The move to Nairobi opened up new research options for me, because it wasn't long before Richard asked if I would prepare some fossils for him. He was in a fix. There were no trained Kenyan preparators in those days, and Ron Clarke, an Englishman who had been preparing fossils for Richard and Louis, had just left to pursue a Ph.D. in paleoanthropology. Before Ron left, he had done a heroic cleaning job on another specimen that became known as *Homo habilis,* a skull from Olduvai that was assigned the number OH 24. The specimen was lovely, creamy-pink-colored bone, but the skull had been squashed flat during the fossilization process. This attribute caused the skull to be nicknamed Twiggy in honor of the emaciated and close-to-two-dimensional fashion model of the same name. Not only was the fossil flattened but the bone was embedded in the extremely hard gray rock of a stromatolite, a ball-shaped rock formation produced by algae. Cleaning and reconstructing Twiggy was a terribly difficult task. Although Ron did a fantastic job, some of the thin bones of the vault had been irrevocably distorted, and their original shape simply could not be restored. However, the reconstruction was good enough to show that the palate was clearly very similar to Cindy, or OH 13, which had also been assigned to *habilis.*

Secretly grateful that Ron, not I, had had the task of cleaning Twiggy, I nonetheless agreed at once to clean some fossils for Richard. I have always loved this soothing work that imposes a slow and careful pace, and it is fascinating to watch the curves and hollows of a bone emerge, sand grain by sand grain, from the matrix. Since one of the great dilemmas of my youth was whether to attend Cambridge and become a paleontologist or to go to art school and become a sculptor, cleaning fossils provides me a taste of the choice I didn't make. I think my headmaster was right when he told me, "You can always be a scientist and do art as a hobby, but you can never be an artist and do science as a hobby," and I have always managed to use my artistic skills to

further my research. Frankly, cleaning fossils is also the best way
I know to learn a specimen intimately. Gradually, Richard asked
for more help with the preparation and then with the descriptions of the specimens; soon we were arguing companionably
over analyses and interpretations that grew into coauthored papers. We developed a strong mutual respect, and in time even
Louis changed his initial low opinion of me and started asking
for my advice.

As persona grata with the Leakeys, I was present when Raymond Dart examined Twiggy, the new *Homo habilis* skull, for
the first time. He and his wife had stopped off in Nairobi to see
Louis and Mary's new finds. Mary had asked me to get out the
fossils and stay around, since Louis was walking with a cane following a hip operation and it wasn't safe for him to carry a
wooden tray with fossils on it. I also knew the organization of
the collection better than Louis did and could help find other
specimens if need be. I put OH 24 out on a cloth-covered table
and stood back while the Darts and Leakeys arranged themselves around the table. Louis watched eagerly while Dart examined the specimen carefully and in total silence. First Dart
simply looked at it, rotating the skull slowly on the table, taking
in the front view, the left view, the back view, and then the right
view. Then he held it up and turned it slowly upside down and
then right side up again, absorbing its general shape and, I suppose, checking for specific anatomical details. All the while Dart
said nothing. Dart was far more familiar than Louis with the
South African australopithecine fossils, especially those from the
cave at Makapan with which PVT had recently coaxed him out
of retirement. It seemed to me that at least twenty minutes
passed while Dart looked, his wife sitting patiently behind him
and Mary sitting opposite him, smoking her cheroot. Louis sat
at the end of the table, shuffling and fidgeting. He had never
needed to look at a fossil for so long without knowing exactly
what he thought of it; he would have made a snap decision about

it long before; indeed, he sometimes seemed to know what a fossil was going to be before he saw a specimen.

Finally Louis banged his cane on the floor in impatience. "Come on, Raymond," he growled, unable to stand the suspense of waiting for the astonishment and praise he hoped was coming. "You've got to tell us what you think. I've got another meeting to go to!"

Dart smiled sweetly and put the fossil down. He looked up with a mischievous expression about the eyes and said only, "Louis, the resemblances are *striking*." I suppressed a grin, picked up the fossil, and left the room quickly on the pretext of putting it away again. I could see that Dart thought OH 24 was the same as his *Australopithecus africanus* from Makapan, or was perhaps its evolved descendant, but he was too canny to say so. Dart's enigmatic response prevented Louis from losing any face and gave him no opening to try to badger Dart into agreeing with him.

The *Homo habilis* situation, already far from simple, grew even more complicated during an expedition of Richard's in 1972, when Bernard Ngeneo picked up some pieces of "bovid" (antelope) skull that everyone else had been ignoring. Ngeneo (his friends call him by his African surname, not his Christian name) was young and only a cook's assistant on that expedition, but he was keen to earn a place on the hominid gang. Every chance he got, he would follow some member of the gang or walk on his own over the nearby exposures, looking for something he probably wouldn't recognize even if he saw it. Others had seen the "bovid" skull and dismissed it, but Ngeneo didn't know enough to ignore these fragments and marked the spot for someone to return and pick them up. He came back to camp, excited that he had found a hominid. The crew followed him eagerly until someone said, "Oh, are those the bovid skull bits?" The high-flying enthusiasm sputtered until they looked at

Ngeneo's find carefully; then it was clear that the fragments were hominid. That was the instantaneous end of Ngeneo's hours of chopping wood and hauling and boiling water as a cook's assistant: he became a full-fledged fossil finder. The find also made Kamoya Kimeu sick at heart, because he had seen the bones twice and had not appreciated their significance.

The amazing skull that this sheer novice discovered was KNM-ER (Kenya National Museum–East Rudolf) 1470. The specimen combined quite a large brain (about 775 cc, compared with an average of 400-550 cc for robust australopithecines and 1,400 cc for modern humans) with a huge, long face. The palate was very large and square in outline, with large, low-cusped premolars at the corners (a feature seen in australopithecines) and sockets for big incisors at the front. The cheek teeth were of only moderate size. Following Richard's lead, most paleoanthropologists call this cranium simply "fourteen-seventy," and 1470 led to the most serious academic argument Richard and I have ever had, a dispute over the nature of humanness.

When Ngeneo found 1470 in 150-odd badly weathered and broken up pieces, Richard was looking forward to giving a presentation at a conference to honor Elliot Smith in London. It would be an important opportunity for Richard to solidify his credentials as a paleoanthropologist—he was only twenty-eight and was still fighting the prejudice that he had no university degrees—and this new skull would be especially valuable. Richard told me about the specimen, saying that they had found another *Australopithecus boisei,* a robust australopithecine like Zinj. Its palate was enormous, Richard said, and strangely there didn't seem to be much postorbital constriction. This is the pronounced narrowing or waisting just behind the browridge that you can see when you look at the crania of most robust australopithecines from the top. Postorbital constriction is caused by the disproportion between their huge faces and relatively small braincases. The implication was that this large-faced individual was fairly brainy by the standards of robust australopithecines, a unique condition that made it seem urgent to piece the fragments of braincase back together. Richard's wife, Meave, his mother, Mary, and I passed the box of fragments back and forth, night after night. The braincase, as we quickly appreciated, was very large indeed; it proved to be 775 cc, larger than the average estimate for the original *Homo habilis* specimens (which had been 670 cc).

As the day for Richard's departure for England approached, we still could not find a good bony connection that would show us how the face was hafted onto the braincase. In any case, he no longer believed 1470 was a robust australopithecine; in his mind, this was an early *Homo*—indeed, the date assigned to 1470 at that time made it the earliest known *Homo,* if that was what the specimen was—and he needed desperately to know how it fitted together. He also needed some photographs of the new skull that he could show at the meeting.

We soon discovered that the real problem with finding the fit was in our minds. There is a photograph taken in the base camp at Koobi Fora, on the eastern shore of Lake Turkana (then called Lake Rudolf), in 1972 that shows how we were all thinking. In it, Richard and Michael Day, a London anatomist who was collaborating with us, are holding up the two main sections of 1470, the large braincase and the face. The face is placed vertically below the braincase, and rather high up too, because we all expected that a big brain meant a small, vertical face. Then one day I watched Meave pick up a long piece of bone, as if to try to fit it onto the braincase just under the browridges. She no sooner took the piece in her fingers than she immediately put it down again, because it was much too long to bother with. But in that instant I realized that it *was* the joining piece. It made the face much longer, and more protruding, than we had ever suspected. I showed her the fit and we glued it in place triumphantly.

When we showed Richard, he couldn't believe the new shape we had produced. He made us undo the glue join to check the fit of the adjacent pieces under the microscope, but there was no denying the bony continuity between the pieces or the long face it produced. He and Ron Clarke took some slides of the specimen for his meeting, orienting the face in a position that emphasized the large brain and advanced, humanlike features of the skull. The schedule was extremely tight so I didn't say much until later, but I thought this was wrong. I wanted to swing the face out at an angle, because to me 1470's large face made it look like a big-brained australopithecine. I knew Richard longed to have a new, oldest *Homo* to show in London, but I just couldn't square that interpretation with its anatomy.

In London, 1470 was indeed hailed as the oldest *Homo,* particularly since Richard's photos showed the face in a near-vertical orientation. Even the stuffiest of the old guard in Britain, Lord Zuckerman, praised Richard, albeit in a characteristically patronizing manner. A stunning photograph taken by Bob Campbell for the cover of *National Geographic* stared out at the public from newspapers and magazines all over the world. For many people, 1470 became the best specimen of *Homo habilis.* There was a sigh of relief that at last everyone knew what the new species was that Louis had named at Olduvai, and the skull was the occasion of a rapprochement between Richard and Louis, who had been somewhat estranged for a while. It was a wonderfully satisfying discovery emotionally, particularly because that proved to be the last year of Louis's life. Scientifically, it still made me acutely uneasy. I was sure Richard's interpretation was wrong.

Not long after the London meeting, there was an international conference in Nairobi, so I took the opportunity to explain my position in detail and call for a change in thinking. 1470 might have a big braincase, but morphologically it was just an australopithecine, as if you had inserted a bicycle pump into the foramen magnum (the large hole where the spinal cord exits the skull) and inflated the braincase like a balloon. Ignoring cranial capacity, the overall shape of the specimen and that huge face grafted onto the braincase were undeniably australopithecine. I also showed that some of the measurements that made 1470 look more like *Homo* were based on a part of the skull that had been distorted during fossilization. I thought the many features of skull architecture and anatomy were more diagnostic—more telling about the biology of this animal—than any single feature like the cranial capacity. Bernard Wood, another British anatomist on the team, took the opposite point of view: 1470 was big-brained and obviously *Homo*. Large brains had always been the hallmark of humanity, and 1470 even met the old Keithian Rubicon. This difference of opinion made for an uneasy relationship between Bernard and me.

After the conference, as Bernard, Michael Day, Richard, and I tried to write a paper announcing and describing 1470 in detail, our differences crystallized. Michael was noncommittal; Bernard and Richard clearly believed 1470 was the earliest *Homo;* I thought the find represented exciting new evidence for brain enlargement among australopithecines. I didn't understand how they could have listened to my paper and still think 1470 was *Homo*. My conviction grew out of my intimate knowledge of the specimen, which I had learned on a millimeter-by-millimeter basis as I cleaned, reconstructed, and described it. I *knew* what the anatomy said, the way a mother knows her child's face. Why couldn't they see what was so obvious to me? I felt, rightly or wrongly, that they were trying to squeeze the anatomy to fit their preconceived theory rather than shaping the theory to fit the anatomy. Ideas are more flexible than bones, I have always found, and when you reverse this situation you are heading into dangerous territory. Our discussion grew heated, and I lost my temper as they tried to persuade me that we should publish the specimen as the oldest member of the genus *Homo*.

"Well," I said, with a coolness I didn't feel, "if it's going to be *Homo* then maybe my name shouldn't be on the paper at all."

"Fine," said Bernard, too quickly.

There was an awkward silence in the room. I stood up abruptly, picked up the fossil, and left the room to put the skull away. My expression must have conveyed my emotions, because Richard apparently thought I was on the verge of leaving the team as well as the room. Even though Bernard was irritated with my stubborn reiteration of my analysis, it was clear to all of us that publishing the initial description of the skull without me—when I had done so much work preparing, gluing, and describing it—would be wrong. I realized almost as soon as I had left the room that I was getting worked up over nothing. Bernard and I were doubtless equally annoyed with each other, but we didn't have to agree on everything. It was entirely possible to publish the description of the fossil and make it clear that some authors thought it was *Homo* and another thought it an australopithecine with a big brain.

I went back in, sat down, and said something offhand like, "Where have we gotten to with this paper?" The others accepted my peacemaking gesture without comment. We acknowledged a difference of interpretation in the paper, carefully separating factual observations from theoretical interpretations to avoid the nasty trap that Louis, John Napier, and Phillip Tobias had fallen into with their original *Homo habilis* paper. Richard told me later, with a small smile, that if I hadn't come back into the room, from his point of view I would have been off the team too. Leaving the team would have meant that I no longer had first access to fossils found on Richard's expeditions, but I could have started organizing my own expeditions again or simply waited for someone else to invite me to join a preexisting team. But true collaborations are based on more than mere compatibility of needs and aims, and field partnerships are among the most demanding of relationships. This incident was a sort of testing point of our collaboration, each of us looking to see whether the bonds of trust, respect, and cooperation would hold against the strain of intellectual disagreement. They did.

Our discussion brought out a fundamental issue. You have to decide: Does brain a *Homo* make? Or does the fundamental biology of the animal—the forces and adaptations that shape the entire body—count for more? I can't believe that sheer brain size defines humanness; all the rest of the cranial anatomy of 1470 says "*Australopithecus*" to me. If brain size is the sole criterion of humanness, then at some brain size threshold, an ape in an animal body suddenly becomes a human in an animal body, who is just waiting for the evolution of its trunk, arms, and legs to "catch up" and become human too. To me what an animal does every day to make a living, what it *can* do every day to make a living, is determined by more than just its mind. The mind has to have a body to work on, and the body has needs and capacities that set limits on what is possible. The idea of a human mind, a sensibility that evolves separately from the body, makes no sense to me. This is why I still don't believe 1470 to be *Homo habilis,* but mine has not been the most popular view. In most anthropologists' minds, 1470 has become the exemplar of *Homo habilis,* usurping the authority of the true type specimen.

The identity of 1470—or, conversely, of *Homo habilis*—is a deep and serious problem. Louis, John, and PVT made OH 7 the type of *Homo habilis* and included behavioral criteria as part of the diagnosis. Nearly all types are unsatisfactory in one way or another unless they comprise a complete skeleton. If, for example, the type is a mandible or a skull, and I find a foot bone or pelvis, how am I to recognize that species? And if the type includes cranial and postcranial bones, the old question that plagued Dubois reemerges: How secure is the assumption that all of the bones come from a single individual? In practice, paleontologists play a complicated kind of linking game, comparing new specimens to the designated types, identifying them as belonging to the same sort of creature as one particular type, and then using any additional features that show up on the new specimen to enlarge the unofficial diagnosis. It is a somewhat complicated syllogism: "If A is like B in one regard, and C is like B in another, is A therefore like C?" Categorization becomes even more complicated when two different species are very similar in some key body part, such as the palate—and that may be the problem with the definition of *Homo habilis*.

Some specimens like Twiggy (OH 24) and a skull known as KNM-ER 1813, found later at Koobi Fora, are called *habilis* because they resemble Cindy (OH 13) in the palate and teeth, and Cindy in turn resembles the original *habilis* specimen in terms of its mandible and teeth. This is fine; you have a simple, workable "palatal definition" of *Homo habilis*. But other specimens like 1470 and a later specimen from East Rudolf called 1590 have different upper teeth or jaws (among other things) from the Twiggy–Cindy–1813–OH 7 group yet are considered *habilis* because of their large cranial capacity. These discrepancies have made for a diverse group of *habilis* specimens that have different sized and shaped teeth, different palates, markedly different faces (for example, 1470's face is about twice as long as the face of 1813 from between the eyes to the bony sockets for the front teeth), and tremendously variable cranial capacities, from just over 500 cc to about 800 cc. The fundamental question is whether all of these specimens are too variable to be grouped into one species. Differences in shape and dimensions within a single species is a phenomenon often attributed to sexual dimorphism, meaning that size and shape varies with sex. Sexual dimorphism is common among primates, with male gorillas weighing twice as much as females, or male baboons having greatly elongated canine teeth and long, doggy faces while females have small canines and much shorter faces. The typical pattern in primates is for sexual dimorphism to make the males bigger and more robust than females, but some species, like chimps, have only a mild degree of sexual dimorphism and gibbons have none. We know that *Homo habilis* had two sexes and so you might expect to see some variability; the complacent view of the *habilis* problem was that this species was more like gorillas than chimps in this regard. I never put much stock in this supposition.

The sexual dimorphism in *Homo habilis* enlarged from "strong" to "incredible" when the next good specimen was found, in 1986. After Mary Leakey had retired from Olduvai, a team including Don Johanson and Tim White had obtained permission from the Tanzanian authorities to start new prospecting and excavation at Olduvai. That their team took over her site was a sore point with Mary, who had fallen out with Don and Tim over the naming and phylogenetic position of *Australopithecus afarensis*. Many of the original *afarensis* fossils came from Laetoli, another site of Mary's, and yet she disagreed with Tim and Don's assessment of the material so violently that she demanded, at the eleventh hour, that her name be removed from the paper that named and described the new species. Disagreements between Richard on the one hand and Tim and Don on the other had only deepened the rift between the Leakey family and a group centered at the University of California and the Institute of Human Origins, then at Berkeley, California. To have those particular scientists move into Olduvai, the place that was her professional (and frequently personal) home for so many years, was galling. But Tim and Don rightly reckoned that the possibilities of the gorge, with its dozens of archaeological and paleontological sites, were far from exhausted. In the first year, to their delight, Tim noticed some badly broken bones in an area not far from one of the dirt roads; this motley collection seemed to include bits of hominid. In among the tidy dung piles left by dik-diks (an exquisite tiny species of antelope), they collected hundreds of tiny fragments of one skeleton. The specimen came to be known as OH 62. It certainly didn't rival the Nariokotome boy or Lucy in completeness and was, in fact, a pretty unimpressive specimen. There were fragments of vault bones, a palate with some very worn teeth, and small segments of both arms and both legs, which gave us hope once again that the *habilis* problems could be straightened out.

OH 62 was a tiny female, smaller even than Lucy, to our surprise. During the summer of 1986 Richard and I first heard a rumor that Don and Tim had found a skeleton at Olduvai. Later, we were told that they were having trouble with the reviewers at *Nature*, the scientific journal, and publication was delayed. Then, unexpectedly, a package arrived for Richard via Federal Express with a return address in Berkeley, California. He opened it to find a Xerox of a corrected proof of a forthcoming article in *Nature* in which Don Johanson, Tim White, and others described the new skeleton of *Homo habilis*. There was no explanatory note or letter with the proof. Richard brought it around to the office where I was working in the museum. "Look at this, Walks," he said. "What do you think it means?"

At first I thought that Don or Tim had made a silly mistake and had mailed the corrected proofs to us while sending a letter of explanation intended for us in another FedEx package to *Nature*. Then I had another idea that seemed more probable, given that at the time there were bad feelings and intense rivalry between Don and Tim on the one hand and Richard on the other. "They probably thought you were the *Nature* reviewer who was giving them a hard time," I speculated. "Maybe this is a message saying that *Nature* is going to take it despite your attempts to block publication." Richard grinned and chuckled a little at the idea of being credited with powers he didn't possess; *no one* could block the publication of an important specimen like OH 62 if it were competently described. We sat together over lunch at a nearby restaurant to go through the paper and see what it meant. When we had read it and peered at the poor copy of a small photograph comparing the diminutive bones of OH 62 with those of Lucy, we were less impressed than we had expected to be.

"We've got lots of partial skeletons more complete than this in the vault," I objected a trifle indignantly. "There are things we've never even bothered to describe in any detail because they weren't good enough."

"Let's go take a look," Richard replied, and we got up from the table and hurried back to the museum. It may seem incredible that we had partial skeletons sitting around that had not yet been properly worked up, but that fact was the result of a deliberate policy we had formulated. Richard wanted to be responsive to long-standing criticisms, aimed largely at his father, that important Kenyan specimens were sequestered for years while the chosen elite studied them with great thoroughness and infinite slowness. In the meantime, many qualified paleoanthropologists who wanted to see the new material were frustrated because it was still "under study." I suggested to Richard that we initiate a policy of publishing the basic description of each fossil hominid as rapidly as possible in a good scientific journal like *Nature* or the *American Journal of Physical Anthropology* and then, after the initial publication, making the specimen available to any scientist who wanted to study it. The analyses by our team could be as lengthy and painstaking as anyone wanted to make them, but they would be published as a series of monographs on topics—the skulls; the postcrania; the paleoecology, or environmental setting; and the paleobiology, or adaptations and lifestyle—that would synthesize material from all the East Turkana remains. This meant that, as long as new homs were being found at a fast pace, the analyses tended to lag behind.

Among our collections we found a partial skeleton that had been numbered KNM-ER 3735. The first piece, the distal, or elbow, end of the humerus, had been found in 1975, a year when there seemed to be many hominid discoveries. Although we knew that the site, in a region known as Area 116, was littered with fossils—mostly fish and crocodile bones—it wasn't until several years later that Kamoya and the gang got around to sieving for the rest of the skeleton, which had been difficult to spot among the other bones. They were working in 116 in 1978, when, back in Nairobi, Richard suddenly went into acute kidney failure. He had been feeling bad for some time, suffering from high blood pressure, but no one outside the immediate family had suspected that the root of the problem was a badly damaged set of kidneys that were giving out. It was a life-threatening situation. As soon as the problem was diagnosed, arrangements were made for him and Meave to fly to England, where dialysis could be performed and the option of obtaining a kidney transplant could be explored. While the medical team fought to stabilize his condition and check his relatives for a potential kidney donor (in the end, his brother Philip provided a kidney) I tried to take care of more mundane business. I flew up to Area 116 in Richard's plane, using a pilot from Kenya Airways who was one of the few bush pilots Richard respected enough to let him fly his plane, and gave the bad news to Kamoya. He closed down the field season and brought everyone back to Nairobi, along with quite a few pieces of 3735 that had turned up in the sieving.

In addition to the original fragment of humerus, the gang had found many pieces of the skull: enough to show it was smallish, with none of the elaborate bony crests of an *Australopithecus boisei* skull like Zinj. (The skull is so distinctive and extreme in robust australopithecines that they are generally the easiest species to identify.) It also lacked the huge jaw joint that was typical of big-brained skulls like 1470, the fossil that I thought was an advanced australopithecine but that everyone else seemed to take as a large *Homo habilis.* In fact, in every part that was preserved, 3735 looked like the skulls classified as the small type of *Homo habilis,* such as 1813 or Cindy. But, significantly, the specimen included parts of the scapula (shoulder blade), the clavicle (collarbone), the radius and ulna from the forearm, two finger bones, and sections of the sacrum, femur, and tibia. None of these bones was complete, but they were good enough to reveal that the original skeleton had been one with strong, big arms and relatively smaller legs. As soon as Mac and I had taken care of winding down the field season, I took Richard and Meave's children, and casts of the new pieces of 3735, to London, hoping that both would lift Richard's spirits. He and Meave were delighted to be reunited with the children, who had been frightened by their father's sudden illness and the disruption of their world, but he was too sick with toxemia to appreciate the fossils. He lay on the sofa in the flat they were living in while I showed him casts of what the gang had turned up. Simply staying awake and paying attention to anything for fifteen minutes taxed his energies to their limit. He tried to smile and mumbled something about being grateful before his head dropped and he drifted off.

Those were dreadful months for all of us who cared for Richard. He was too sick to think or do anything and Meave bore alone most of the strain of hospitals, children's fears, financial worries, and medical decisions. I cannot describe the feeling when he finally got the transplanted kidney from his brother and could go into the field again, alive and alert once more. He returned with an infectious joie de vivre. But, with all of the emotional upheaval involved with Richard's illness, 3735 had been returned to its case in the "chapel" once again without anyone's ever having examined it carefully.

Now the publication of Don and Tim's new Olduvai skeleton, OH 62, spurred us into doing what we should have done in the first place. Working with two graduate students from Johns

Hopkins Medical School I set out to analyze the limb proportions and body build of 3735, which we compared with data
from a chimpanzee, Lucy, and a human. This series crudely re-
produced a phylogenetic sequence, with the chimp representing
the original, apelike condition, Lucy representing the first evolutionary step on the hominid lineage, and humans representing
the modern anatomy of our lineage. *Homo habilis,* which was
an evolutionary link between Lucy and humans, could be expected to be more humanlike than Lucy with an anatomy less
dominated by strong, long forearms and more dependent on
hindlimbs.

I use the word *hindlimb* rather than *leg* deliberately. Hindlimb is a more accurate term because, to an anatomist, the leg
refers only to the lower part of the hindlimb, while the thigh is
the upper part. Besides, *hindlimb* and *forelimb* emphasize the
continuity in structure between these hominids and other animals. It is important to embed ourselves and our ancestors
firmly within the mammalian world; keeping the same termi-
nology for humans and other mammals is one antidote to the potential perils of homocentrism. We are first and foremost
mammals and have to operate under the same biological, physiological, and biomechanical "rules" as any other mammal.
There was no special creation for humans, and there are no special exemptions from the constraints of anatomy.

Although our comparisons were limited by the incomplete-
ness of the fossil material, the answer was clearly not as expected.
The bones of the arm and hand—scapula, radius, ulna, and finger bones—were all strong and robustly built. The scapula was
large, with thick bony crests to demarcate the attachment sites of
hefty shoulder and upper arm muscles, much like the chimpanzee that we examined. Moving down the arm, the bony areas
where the muscles attached to flex the elbow showed that this
action was more powerful than in a chimpanzee and much

stronger than in a human. The finger bones, similarly, showed evidence of an impressive grasping ability comparable to the chimpanzee's. All of these features and measurements pointed to considerable climbing ability and heavily muscular arms, and in every measurement Homo 3735 was markedly larger and more strongly built than Lucy. However, the hindlimb told a different story. The sacrum closely resembled Lucy's in both size and morphology but was small compared to a chimpanzee's. The femur and tibia of 3735 were only slightly bigger than Lucy's, which was in turn about the size of a chimpanzee's; the largest hindlimb dimensions, by a substantial margin, were those of the human.

In other words, 3735 had a Lucy-like, chimp-size hindlimb, but its forelimb was even bigger, longer, and stronger than Lucy's: more like the chimp's. Rather than representing an intermediate between Lucy and humans, 3735 looked very much like an intermediate between the ancestral chimplike condition and Lucy. Our findings were rather surprising, but in one respect they were predictable. Our analysis of the forelimb of 3735 offered good support for a claim made a few years earlier by Randall Susman and Jack Stern, two anatomists at the State University of New York at Stony Brook. Before OH 62 was found, they had examined the isolated postcranial bones attributed to *Homo habilis* from Olduvai and concluded that the species probably combined bipedalism with agile tree-climbing. They viewed the curved fingers and toes, strong arms, and flexible shoulders of *Homo habilis* and other early hominids as arboreal or tree-climbing adaptations. Owen Lovejoy, whose work had established that Lucy was a well-adapted biped, and others on Don Johanson's team hotly contested the Susman-Stern claim. The proponents of full bipedalism argued that these adaptations had evolved because of an arboreal life-style in the ancestry of all hominids; these traits were retained as evolutionary "holdovers" in *Homo habilis* and other hominids because of their usefulness in terrestrial tasks requiring arm strength and manipulation. The question of whether early hominids were fully committed to bipedalism, or whether they combined bipedalism with habitual tree-climbing into a unique locomotor pattern unknown today, is still one of the great debates of paleo-anthropology. All we could add to the debate was the observation that our partial skeleton of 3735 seemed to show the same sort of powerful development and forelimb dominance that an arboreal species needs.

Another question was whether these adaptations were also present in OH 62. We couldn't measure OH 62 or analyze it in any detail based on the information in the *Nature* note since it was, by the journal's dictates, only about a thousand words long, but we suspected that its proportions were unusual. One mea-sure of the relative sizes of arm and leg that paleoanthropologists often calculate is the ratio of the length of the humerus to the femur in a single individual; this ratio is also a good indicator of locomotor pattern. Primates that habitually climb in the trees have longer arms than legs, while bipeds and leaping primates have longer legs than arms. For example, chimps have a ratio of 100 percent or slightly more, meaning that their arms and legs are nearly equal in length; the extremely arboreal gibbon has a ratio of 132 percent. Earthbound humans have a ratio of 70 per-cent. (Unfortunately, we couldn't take the measurements needed to calculate the ratio on 3735, because its bones were too incomplete.) The *Nature* paper said that OH 62 had a ratio that was close to 95 percent: only a little more human than a chimp's ratio. Even Lucy, a representative of a species ancestral to (and thus closer to apes than) OH 62, was more human than OH 62 in this regard. Although both petite and long-armed, Lucy's ratio was estimated at 85 percent. Sometimes the new partial skeleton from Olduvai was dubbed "Lucy's little sister" or "Lucy's child," to indicate its descent from a Lucy-like ancestor, but this long-armed creature certainly didn't look more human than Lucy. The family resemblance was, to say the least, strained. Indeed, it was apparent from the photograph in *Nature* that OH 62 was even tinier than Lucy, who herself had stood only about three feet six inches. None of these body proportions suggested a species that I would happily consider the female to a 1470-like male (which many considered a typical *habilis* male).

Why, with this apelike build, was OH 62 identified as *Homo habilis*? As Tim and Don explained in the *Nature* paper, the palate and teeth of OH 62 are similar to those previously identified as small habilines, especially Twiggy (OH 24) and a skull from South Africa known as StW 53 (which became an important specimen in a later piece of research). It was a dental hominid and a postcranial but upright ape.

It was only in 1991 that the first detailed analysis of the OH 62 limb bones appeared. It was published in the *Journal of Human Evolution*; the research had been carried out by Sigrid Hartwig-Scherer, a young Swiss anthropologist working with her former adviser, Bob Martin, an old friend of mine. As we had done with 3735, they compared OH 62 to a chimpanzee, to Lucy, and to modern humans in terms of the features and proportions of the limbs. This was a way of asking where OH 62 belonged on the continuum that started with an ape, included a rather apelike biped (Lucy), and ended with a fully modern biped (human).

In every measurement Sigrid and Bob could take, OH 62 was even more apelike than Lucy. Their results confirmed ours and produced a puzzling dilemma. In addition to these two partial skeletons of *Homo habilis,* we had a number of large, isolated limb bones ("isolated" in the sense that we hadn't recovered any other bones from the same skeletons) that we had always thought were probably *habilis* bones. Our tentative identification of these bones as *habilis* had been based on two facts. First, they were the right age to be *Homo habilis* (between about 2 million and 1.8 million years old). Second, they differed morphologically from the bones of a partial skeleton of *Australopithecus boisei* in our collection. Theoretically, any non-*erectus,* non-*boisei* hominid from that time period simply had to be *habilis,* because that was the only other hominid known to exist in the area. Because they were large, we had assumed them to be from males like the one represented by what most people saw as large habilines, like the big-brained 1470 skull.

However, these isolated bones are not apelike like the bones from 3735 or OH 62. An attempt to synthesize all of these data into one coherent explanation suggests a nonsensical evolutionary pathway: first our ancestors got less apelike, then the females got *more* apelike while the males continued to get less apelike, then the females reversed evolutionary directions and both sexes continued to follow the trajectory toward decreasing apishness (or, from the other perspective, increasing humanness). The alternative interpretations are either that these long-armed, partial skeletons weren't *habilis* at all or, if they were, then *habilis* wasn't an intermediate between australopithecines and *Homo erectus.* And, if there is one attribute that has defined *habilis* from the very beginning, it is its intermediate position between australopithecines and *Homo erectus.* That is why I don't like *habilis* as a species; something is all wrong with it and always has been. I don't think it is disloyal to my old adviser, John Napier, to say that he got it wrong (for once) when he participated in naming and defining *Homo habilis.*

If *Homo habilis* had had a troubled sex life before this, now the problem was acute. The faces, the brains, and now the bodies too were too different for a male-female difference. You can make males and females out of the less extreme variability among the small habilines, but then you have *three* sexes: a small female, a small male, and a large male. If you don't like that argument, then you can make the "big male" 1470 (with its capacious skull but australopithecine face) a different, novel species of which we had a representative of only *one* sex. This hypothesis is more plausible than the three-sex hypothesis, I think, because we have precious few fossils of 1470-like animals anyway. Perhaps when we find a few more we can restore 1470 to a two-sex condition. I think he will be much happier that way. But, ironically, this muddle leaves *Homo erectus* without a clear ancestor, without a past. *Erectus* may be the now-found missing link, but the link to which it was connected is now a missing one as well.

7

THE SINGLE SPECIES SOLO

There were other questions in the 1960s and 1970s that seemed more preoccupying than those centering on *Homo erectus* or even *Homo habilis*. As new specimens were found almost daily, *Australopithecus* rapidly became the missing link that everyone most wanted to find. Soon I, too, was fascinated by its primitiveness, by its combination of human and apelike traits. Working on the ever-growing number of australopithecine specimens that Kamoya and Richard were finding at East Turkana prolonged the spirit of my days as a graduate student with John Napier; here were the most exciting new finds, and I was part of the team that analyzed and interpreted them. It may seem unrelated, but our struggles to define and understand these earlier hominids had a lot to do with how we saw *Homo erectus* years later, when we finally found the Nariokotome boy, for they established the nature of our entire zoological family, the Hominidae. We had to understand the family first before the individual species could make much sense.

The question of how many early hominid species there were was one of the first issues begging for resolution. Dart, Broom, and Leakey had set up a pantheon of genera: *Zinjanthropus, Paranthropus, Australopithecus, Telanthropus,* and *Plesianthropus,* not to mention *Homo habilis, Homo erectus,* and all the old names for *erectus,* like *Sinanthropus* and *Pithecanthropus.* Into these categories were placed a plethora of specimens known by nicknames and the new horde of fossils that were called by their museum numbers. It was impossibly confusing trying to draw a phylogeny that represented the evolutionary relationships among all of these myriad species, especially since many of the bearers of different names were so similar physically that they were surely one species. The first round of problems involved the australopithecines; more advanced forms, like *Telanthropus* and *Homo habilis,* were more or less ignored, while all the various old names for *erectus* were formally abolished but occasionally still used in colloquial contexts.

A conservative view was that there were two types of australopithecines, little ones and big ones, also known as *Australopithecus africanus* and *Australopithecus robustus* (the latter was formerly called *Paranthropus robustus* by Broom). The Taung baby and Broom's specimens from Sterkfontein (*Plesianthropus*) would go into the smaller and less sturdy species, *A. africanus,* while Broom's specimens from Kromdraai and Swartkrans were the larger, more robust australopithecines. Zinj and similar specimens from East Africa were obviously similar to the South African robusts, though their skulls are so strongly built as to be considered "hyperrobust," and I would place them in the same genus but in their own species (*Australopithecus boisei*).

Another possibility for the australopithecines was that the robusts were the males and the graciles were the females of a single species, *Australopithecus africanus.* (They would have to be *africanus,* not *robustus,* because Dart's was the older name and priority always takes precedence in taxonomy.) But if these specimens were just males and females of the same species, then you had to postulate a very peculiar social arrangement, for all the males came from Swartkrans while all the females were found across the valley, a mile or so away, in Sterkfontein cave.

The hypotheses about australopithecines can be diagramed as two alternative phylogenetic trees, only one of which can be correct.

At times, the debate seemed so nonsensical as to make it tempting to throw darts at the two diagrams blindfolded as a means of choosing between them. A more scientific approach to testing the hypotheses was to examine the sexual dimorphism statistically, seeing how much males and females of living species differed in such features as tooth size, cranial capacity, browridge dimensions, inferred body size, and so on, and then comparing this degree of dimorphism to that found among the fossils. Statistics are slippery things, however, beloved by people who generally don't "see" shape very well but understand numbers. Statistics always involve assumptions about the nature of the underlying reality from which you have selected a sample; if those assumptions are incorrect, they can warp the results into something that is biologically meaningless. For me the anatomy and morphology were quite clear: gracile and robust australopithecines were different sorts of animals from each other, and early *Homo* was a third type.

It was a complex tangle, particularly since the lack of volcanic layers in South Africa meant that potassium-argon dating was inapplicable. There was, and still is, no good way to get an accurate date on the South African cave sites, and their stratigraphy is too complex for easy correlations to be made from one cave to another. You can't be sure who came first or even if any one of them came first, for they might all have lived at the same time.

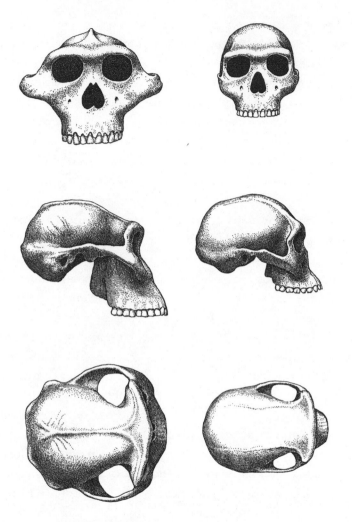

Sex or species differences? In the 1970s, anthropologists argued over the number of hominid species that lived between 2.5 million and 1.5 million years ago. Two skulls from this time period in Kenya are shown: (left) a robust hominid, *Australopithecus boisei* (KNM-ER 406), and (right) a gracile hominid that some people think is *Homo habilis* (KNM-ER 1813). The single species hypothesis postulated that these were males and females of a single australopithecine species with substantial sexual dimorphism. I think that these are two distinct species with less marked male-female differences.

Even if the dates were a muddle, it seemed unlikely that only one unbranching evolutionary lineage could be cobbled together from the specimens we already had. There had to be some side-branches, evolutionary experiments in being a hominid that went extinct without issue.

The argument sputtered along inconclusively for years. John Robinson, who had begun his career as an assistant to the eccentric Robert Broom, continued to excavate and study South African hominids after Broom's death. Robinson hypothesized that there was an ecological difference between the two distinct lineages that he saw in the fossils. One of these lineages was the robust australopithecine group, with its cresty heads, flat or even dished (concave) faces, and large teeth. The other was the more gracile lineage, made up of small australopithecines that, in Robinson's view, evolved into *Telanthropus*, a more *Homo*-like hominid. The gracile lineage featured more globular braincases, smaller cheek teeth, larger front teeth than robusts, and a more protruding face. What distinguished Robinson's interpretation was that he identified the primary distinction between the two lineages as dietary, with the smaller-toothed lineage leading to *Telanthropus* and *Homo* eating a more omnivorous diet and the larger-toothed robusts eating a low-quality, vegetarian diet. The huge teeth and powerful chewing muscles were adaptations to grinding up great quantities of low-quality fibrous food. Both australopithecines were upright bipeds, and Robinson thought that the robusts had probably evolved elsewhere and migrated into the other lineage's territory.

This hypothesis had two big advantages. First, it *explained* the morphological differences that could be observed on the specimens, rather than just documenting them or drawing phylogenetic trees to link them up in an evolutionary sequence. Second, it transformed the arguments about different hominid species from one based on features and inferred evolutionary relationships to one based on attempts to reconstruct and understand the biology and adaptations of our ancestors. This made the homs more than a list of names or traits; it started to bring them to life.

Argue as he might, John's ecological hypothesis was not universally accepted. The *Telanthropus* material was sadly fragmentary. His assertion—that its molar teeth were not only much smaller than those of robusts, they were also morphologically different from them—failed to convince everyone. There is an irony in this, for, with our much better knowledge of the early hominid record today, a single molar of a robust is now regarded as being virtually unmistakable. If *Telanthropus* had been a robust, you would think everyone would have known it. John was right about *Telanthropus,* but he did not carry the day. His intensity and singleness of focus made for heated attacks and rebuttals in the literature.

Among the main antagonists to John's dietary hypothesis were C. Loring Brace and Milford Wolpoff, two anthropologists at the University of Michigan who proposed a strong theoretical reason why there should be only one species of hominid during any given time span. The genesis of the idea was in a seminal paper by Loring about Neandertals, not australopithecines. He believed, and said, that culture was a primary means by which humans adapted to their environment. By *culture* he meant any sort of learned, structured behavior. Evidence of culture thus became anything created behaviorally and regularly, such as tools, shelters, or altered natural objects (sticks, stones, or bones) that attested to the human modification or manipulation of the environment and its contents. What distinguished the human lineage was *extra-somatic* (outside the body) evolution. Thus, for example, rather than evolving sharp teeth to cut meat, hominids learned how to make "teeth" by hitting rocks together to make sharp-edged tools; rather than evolving thicker fur or body hair to keep warm in temperate climates, hominids learned how to control fire and how to make clothing and shelters. (These examples sound rather teleological, as if our ancestors decided to skip evolving teeth and make tools instead, although I do not mean to imply this.) In ecological theory, this made culture our ecological niche, and our lineage was the only one that, as far as anyone knew, had ever been able to occupy this niche. We were indeed a unique evolutionary product, one with an ancient, and troubling, penchant for changing our environment rather than ourselves.

Milford Wolpoff, then still a student and an old friend of mine, took this idea one crucial step further. Milford thinks broadly, and argues loudly, and he saw an important implication in Loring's notion that culture was the hominid ecological niche. He knew that one of the axioms of ecology is that two closely related species cannot occupy the same ecological niche in the same place at the same time. Sooner or later, one of the two species will outcompete the other. As a consequence, the less well adapted species, what ecologists call the less "fit" species— although they are using "fitness" in a special sense—will be driven into local or permanent extinction. When Milford applied the principle of competitive exclusion to early hominid evolution (not Neandertals), he came up with the single-species hypothesis. In a wonderfully thought-provoking paper, he explained his idea:

> Because of cultural adaptation, all hominid species occupy the same extremely broad, adaptive niche. . . . Although culture may have arisen as a defensive survival mechanism, once present, it opened up a whole new range of environmental resources. Culture acts to multiply, rather than to restrict, the number of usable environmental resources. Because of this hominid adaptive characteristic, it is unlikely that different hominid species could have been maintained.

In short, *if* culture were the hominid niche, then there could *never* be more than one species of hominid in any one region at any point in time. Milford believed John Robinson's differentiation between *Telanthropus* and the robusts thus had to be in error, and said so in print. It was an audacious move for a graduate student who had, a few short years before, spent a semester as a visiting student with the well-known John Robinson. Milford's idea was attractively grounded in ecological theory, but so was John's. Starting in 1968, Milford explained and elaborated his counterthesis. There was a lovely logical clarity about it, and if you missed it the first time he was always ready to explain it again. Many anthropologists began to wonder if robust and gracile australopithecines were just males and females of the same species. Stone tools, the earliest and most durable form of culture, apparently appeared along with the earliest hominids, so the fossil record seemed to provide direct evidence for the hypothesis. (It was only later that even earlier australopithecines were found, showing clearly that bipedal hominids predated the appearance of stone tools by more than a million years.)

The beauty of the single species hypothesis was that it was testable. If it could be demonstrated that two distinct hominid species coexisted sympatrically (living in the same place), then the theory would be falsified. But as long as all differences among hominids could be cast as sexual dimorphism, the theory was supported. John tried but was unable to demolish Milford's theory using the *Telanthropus* material from Swartkrans, which simply wasn't sufficiently convincing.

The obvious place to test the hypothesis stringently was Koobi Fora, the region east of Lake Turkana (then still called Lake Rudolf), where Richard and Kamoya had been working since the late 1960s. The hominid specimens were good, fairly complete, and the team had good stratigraphic control over their placement, so, as fossils continued to accumulate, we would be able to determine if two different homs ever lived in the same place at the same time. In fact, the very first skull of *Homo erectus* that was found at East Turkana finally settled the single species debate.

The trick was, of course, to find two sympatric hominids that were clearly different species. The Zinj skull that Louis and Mary had found at Olduvai in 1959 showed extreme and unmistakable anatomical adaptations; in 1969 Richard recovered a similar skull, KNM-ER 406, clearly the same species. It was one of the few skulls he ever found himself. Richard refused to give it a nickname, as his parents had always done with their fossils, and simply called it "four-oh-six" for its catalogue number. It was the triumph of one of his first field seasons, which was conducted on camelback with his wife, Meave, and Kamoya by his side. The photographs look like a romantic adventure out of an Indiana Jones movie, but even Kamoya complains that the camels were hopelessly uncomfortable, and bad tempered too.

The specimen known as 406 is a beautiful, hyperrobust cranium. As the hominid gang grumbled good-naturedly, envious that they hadn't spotted it first, it is a big enough specimen for even a *mzungu* (white person) to find. Its dished face, large crest, and massive palate said clearly that it was an *Australopithecus boisei,* even though its teeth had broken off during the fossilization process. There were still enormous tooth roots, counteracting any speculation that small teeth would fit in its mouth. Besides, we had several huge jaws full of teeth that fit well enough on 406 to show what its jaws and teeth must have been like. To disprove Milford's hypothesis we needed a second species of hominid, which we had, but it couldn't fulfill the second criterion of contemporaneity with 406.

It wasn't until 1975 that Bernard Ngeneo made a wonderful find that brought the single species debate to a close. It was the skull numbered KNM-ER 3733, known as "three-seven-three-three." Cleaning and reconstructing 3733 was a serious problem. The face was almost complete, once I had fitted the fragments back together, but it was fragile. The bones were thin, as you might expect with a more gracile, humanlike face. On the other hand, the braincase was sturdy, solid, and completely filled with stony matrix the consistency of cement. I knew if I glued the face onto the braincase, the face would get crushed by the weight of the braincase the first time someone put the specimen down on a table. Richard had grandiose ideas about cutting into the braincase with lasers or special saws, but I talked him out of it. "Any way we cut into it," I argued, "we're going to lose some material. Let me open it up by propagating the natural cracks between the fragments. You'll never know it's been done and we'll lose nothing." Richard agreed and left town: he couldn't bear to see me do it.

I started by pasting cigarette papers all over the surface of the braincase with Gloy, a water-soluble glue rather like library paste. When the fossil cracked, the thin papers would hold all the tiny pieces in place so they could be reassembled. The trick was to use Gloy for this step and not Durofix, the acetone-based glue we had used initially to glue the fossil fragments to each other. This meant that I would be able to remove the papers, by dissolving the Gloy with water, without simultaneously dissolving all the Durofix that held the skull together. Once I had papered the skull, I placed it upside down on a sandbag and took it out on the veranda of the small stone house on the museum grounds where I was working at the time. I had a small audience of anthropologists and gang members.

I put a nine-inch-long cold chisel into the foramen magnum and hit it with a small sledgehammer. There was an awful *whang* and nothing happened.

I hit it again, harder. Nothing happened.

Still harder: no change.

Tim White, who was then a graduate student at the University of Michigan, was watching this historic moment and steadying the skull. Like everyone else, he was probably wondering if I was going to smash this important fossil into tiny bits and if Richard would have my guts for garters as a consequence. No one had ever done anything like this before.

"I don't think you're hitting it hard enough," Tim offered.

I looked at him and laid the cold chisel against the top of the wall of the veranda, which was made out of a hard volcanic ash known as Nairobi ignimbrite. I hit the wall with the same force and a huge piece of rock spalled off.

"Oh," Tim said.

I sent Kamoya off to get a new, sharp cold chisel. I tried again, and again, hitting the chisel as hard as I could.

After a dozen blows, the braincase cracked open neatly, just as I had predicted. Then it was just a matter of removing the rock inside using an air scribe, a sort of miniature jackhammer, under a binocular microscope, so that I could always see the distinction between matrix and fossilized bone. Once the matrix was gone, I glued the cranium together again, removed the cigarette papers, and attached the face. By the time Richard got back into town, the specimen was beautiful. We never talked about the process, but he did complain about the damaged veranda.

3733 was an exquisite *Homo erectus* skull that recent work has dated to 1.8 million years, making it probably the oldest known specimen of *erectus* anywhere. It looked startlingly like the casts of the composite Peking Man skull that Weidenreich had modeled up from several finds at Zhoukoudian. (To my eyes, it also was a dead ringer for John Robinson's more fragmentary *Telanthropus* in South Africa, making clear what *that* was.) But the most important point was that 3733 was found at the same stratigraphic level (above the volcanic ash known as the KBS tuff) as Richard's skull 406, which was clearly *Australopithecus boisei*. Whatever the absolute dates were, these crania were contemporaneous and sympatric, the two conditions under which competitive exclusion ought to work. Richard and I knew we had disproved Milford's hypothesis as soon as we saw 3733; it couldn't possibly be the same species as 406. We wrote a draft of a paper for *Nature* called "*Australopithecus, Homo erectus,* and the single species hypothesis" before I returned to the States.

I did not want to blindside Milford, so I called him in Michigan. "I've got something to show you, Milford," I said. "New fossils." It was irresistible bait, and he arranged for me to give a seminar to the faculty and graduate students at Michigan. For most of the hour, I reviewed the East Turkana fossil record—jaws, teeth, postcranial bits. Then, at the end of my lecture, I simply said I was going to show them the newest material from Koobi Fora. I hadn't even had time to develop my slides, so I had to put Polaroid snapshots of 406 and 3733 into an antique epidiascope (one of those gadgets that works like an overhead projector except it does not require a transparency) that somebody unearthed from a closet somewhere. While all eyes were focused on the screen, I explained the identical stratigraphic positions of the two skulls. Then I waited. Everyone in the room knew enough to recognize *Homo erectus* when they saw it, and they did.

The lights came up and all the students were blinking, as much from mind bending as from the change to brightness. They thought about what I had shown them: two skulls, one with a small brain and huge crest (406), one with a big brain and no crest (3733); two skulls, one of a species with teeth so large that four or five of the other's teeth would fit on each of its crowns; two skulls, one an eating machine and the other a thinking machine; two skulls, one time horizon. Did this mean that the single species hypothesis was dead? There was a stunned silence.

Milford, ever the devil's advocate, immediately questioned the stratigraphy, trying to find an error that would allow his theory to survive. He had had a good idea, an insightful perception, and it was tough to relinquish it. For many years, I thought Milford had admitted that his pet hypothesis was proven wrong, but an idea you cherish is sometimes more compelling than a lover. I was amused to read a paper of Milford's in 1993 in which the "replacement" of *Australopithecus boisei* (things like 406) by *Homo* (things like 3733) is described as "the best-documented case of competitive exclusion" in our lineage, even though it "took place over a period of more than one million years, measured by the unequivocal coexistence of at least two different species of hominids." It is a poor sort of competition that takes over fifty thousand generations to have any exclusionary effect. I would have said this was the best-documented case of nonreplacement that strongly suggested no meaningful competition between the two species.

The question remained: *Why* was it wrong? Facts and theory, said Goethe, are natural enemies. So it was with the single species hypothesis. The only "if" in the theory, the only proposition that must now be in error, was the fundamental idea that culture was the ecological niche of the Hominidae, the zoological family to which we belong. We are surely culture-bearing animals, and undoubtedly culture is a major means by which we adapt to our environment. It just wasn't the ecological niche of early hominids. It can't have been or 406 and 3733 and their kin wouldn't have been sympatric and contemporaneous for so long.

Once we knew that there were at least two hominids alive at once, then it was possible to conceive of several contemporaneous and sympatric species: no one doubted any longer that *Australopithecus africanus* and *Australopithecus robustus* had coexisted in South Africa as separate species. As Richard and I said in our *Nature* paper, "The new data show that the simplest hypothesis concerning early human evolution is incorrect and that more complex models must be devised." Complexity is rarely lacking in paleoanthropology.

While the single species hypothesis was being tested and found wanting, a team led by Don Johanson and Maurice Taieb had been working in the Afar region of Ethiopia. They soon found the famous partial skeleton known as Lucy, which they named *Australopithecus afarensis* in 1978 and which was then believed to be the earliest and most primitive hominid. The discoverers argued that Lucy and her conspecifics (others of the

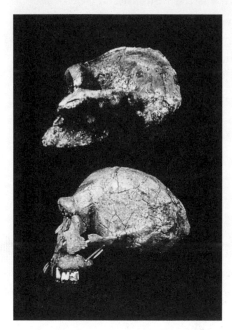

The eating machine versus the thinking machine. Comparing the crania of *A. boisei* and *H. erectus* in top and side views highlights their differences in adaptation. The *H. erectus* specimen (KNM-ER 3733) is on the right in the top view; *A. boisei* (KNM-ER 406) is on the top in the side view. The enormous teeth and jaws of *A. boisei* were driven by massive chewing muscles, which were anchored to striking bony crests at the top and back of the skull and to its strong cheekbones. Its heavy face is re-inforced to withstand chewing stresses. In contrast, the chewing muscles of *Homo erectus* were relatively small and its cranium is dominated by the large, globular braincase, which housed a brain almost twice as big as that in *A. boisei*. The coexistence of these two species at Koobi Fora demolished the single species hypothesis.

same species) were ancestral to all later hominids. This find, wonderful though it was, still left open several questions about the relationship of *Australopithecus* to *Homo.* Which australopithecine, *afarensis* or *africanus,* gave rise to *Homo*? Did gracile australopithecines evolve into the robusts or did the latter arise separately from *afarensis*? These issues are still being debated. My favorite hypothesis at present is shown in the next figure, but new finds always have the potential to revise any phylogeny completely.

At least one possibility, the single species hypothesis, had been resoundingly eliminated. This engendered a tremendous new emphasis on ecology in thinking about all of these hominids, who disconcertingly seemed to have been running around together. How were they differentiated and how did they coexist? Now John Robinson's hypothesis reemerged. He had postulated that the gracile lineage, leading from *Australopithecus africanus* to *Homo,* was more carnivorous, while the robust lineage headed toward extinction as a strict herbivore or vegetarian. He supported this by pointing to the larger molar teeth of the robust lineage—giving a bigger grinding area for chopping up vegetation, like a horse's or an antelope's teeth—and the smaller molars and slightly more elongated canine teeth of the gracile australopithecines.

For reasons of health, John dropped out of the academic argument in the early 1970s. He had had a severe heart attack, and his doctor gave him strict orders not to get excited and overwrought. At the last conference I attended with John, in 1973, he prefaced his paper by repeating his doctor's instructions and announcing that, although he would read his paper, he would not answer questions or discuss it lest he become upset. It was an effective strategy for quashing dispute.

If John was not going to pursue his ecological hypothesis, it was up for grabs, and I knew how to test it. I had a strong intuition that different foods would leave different types of wear on

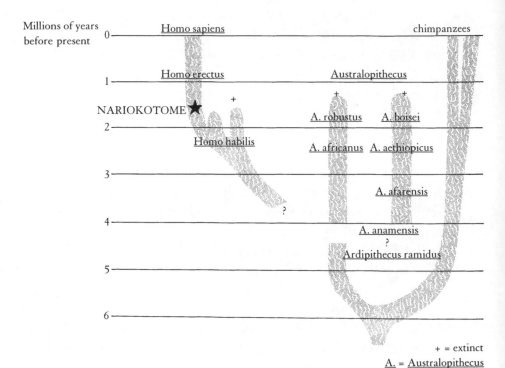

Millions of years before present

Homo sapiens   chimpanzees

0 —

1 —   Homo erectus   Australopithecus

NARIOKOTOME ★

+

A. robustus   A. boisei

2 —   Homo habilis

A. africanus   A. aethiopicus

3 —

A. afarensis

?

4 —

A. anamensis

?

Ardipithecus ramidus

5 —

6 —

+ = extinct
A. = Australopithecus

More fossils mean more possibilities. Drawing a family tree for the Hominidae becomes increasingly difficult as species are added. In 1978, anthropologists named *Australopithecus afarensis*, which lived between about 2.8 million and 3.6 million years ago and whose best-known representative is the partial skeleton known as Lucy. On a recent expedition that Meave Leakey led, the hominid gang found specimens from Kenya that are yet another species of fossil hominid dated to about 4.1 million years. Meave and I have named it *Australopithecus anamensis,* from the Turkana word *anam*, meaning "lake." The new specimen is bipedal and shows a curious mixture of features, some primitive and *afarensis*-like and others more advanced. Even older (about 4.4 million years) is *Ardipithecus ramidus,* a new hominid named by Tim White's team in 1995.

an animal's teeth, at least at a microscopic level. If this was true, then all I had to do was study living animals of known diet, characterize the dental microwear, and compare the unknowns (the fossil forms) to my matrix or reference set of knowns. Then I could deduce the diet of the fossil species. What I needed was a really good test case: two animals of similar size and tooth structure, living in the same habitat and eating different foods. Heindrick (Enrico) Hoeck, a friend who was an ecologist working in East Africa, provided the perfect opportunity.

Enrico had been studying hyraxes, a particularly engaging little animal found over much of Africa. In general size and appearance, hyraxes look rather like groundhogs (marmots) or fat prairie dogs, but their teeth resemble miniature rhinoceros teeth and on their feet they sport soft little leathery pads, like pint-size elephants. Belying their charming appearance, some hyraxes call to each other at night with a bloodcurdling sound, like that of a woman being strangled. Improbably, elephants were once considered to be hyraxes' closest living relatives. Since one is small, furry, tuskless, and trunkless while the other is enormous, pachydermatous, tusked, and trunked, you can see the relationship is not a close one, and just who hyraxes' relatives are is still a much-debated topic.

Enrico had spent several years observing a single kopje, or rocky outcrop, in the middle of the Serengeti Plain in northern Tanzania. These kopjes rise well above the plains, rather like an island in the ocean, and support a different community of animals from the zebras, wildebeests, and gazelles that live on the plains. Kopjes are full of rocky crevices and caves and often sport small trees or bushes. On Enrico's particular kopje there were two species of hyraxes so different that they are put into two different genera. *Procavia johnstoni*, the rock hyrax, is slightly larger, with redder fur. *Heterohyrax brucei*, Bruce's hyrax, is smaller and browner. But here they were, living so intimately that they shared the same dens, defecated in the same latrine areas, suckled each others' young, and huddled together for warmth on cold nights in one great hyrax ball. The difference was dietary. In the dry season Bruce's hyraxes spent their days climbing into the bushes and trees on the kopje and eating leaves, so that 92 percent of their diet was browse; while the rock hyraxes split their diet more evenly between browse (57 percent) and grass (43 percent). In the wet season Bruce's hyraxes continued to browse on leaves (81 percent), but also ate a little grass, or grazed (19 percent); while the rock hyraxes went to the other extreme and browsed for only 22 percent of their diet, clambering down off the kopje to graze for 78 percent. In short, Bruce's hyraxes were much more committed to browsing, while rock hyraxes favored a larger percentage of grazing. During the course of his study, Enrico collected nineteen rock hyrax and twenty-three Bruce's hyrax skulls from individuals who had died of natural causes. This meant we had specimens from individuals whose dietary habits he had observed directly. Enrico sent me the skulls, and my technician, Linda Perez, and I prepared them for our study. What we wanted to do was look at them under the scanning electron microscope, or SEM.

In principle, SEMs work like an ordinary light microscope, except that the SEM shows you an image created by electrons, which are bounced off the object in question, rather than one made by photons, or light particles. In order to subject the teeth to a stream of electrons, we had to be able to put them into a vacuum chamber. Large specimens simply wouldn't fit into the chamber, and because we hoped to apply this technique to fossil hominid teeth, we had to devise a way to make a faithful replica or cast of the teeth. (I didn't need to think hard to know what Richard would say if I told him I wanted to break the teeth out of his hominid jaws to bring them home and put them into an SEM.) Linda experimented with various brands of silicone-based goo manufactured for dentists who wanted to make dental impressions; one of them proved ideal. It was called Xantopren, a blue substance that came in a tube like colored toothpaste. You mixed it with a catalyst and the compound would set up, or harden, into a rubbery substance in about six minutes. In the meantime, you could apply it to the teeth you wished to examine. When you peeled off the hardened Xantopren, the surface of the mold recorded in extraordinary detail every bump, groove, scratch, and pit on the tooth's surface. We poured epoxy resin into the mold to produce a positive (looking at negatives is extremely confusing visually) that was inserted into the SEM's vacuum chamber, after it was coated with a microscopically thin layer of gold so that it would conduct electrons. With a little practice, we found we could produce replicas so good that we could resolve tiny structures much smaller than the size of a single cell.

Using the SEM is like entering a whole new world. You can manipulate objects in three dimensions—up/down, left/right, and nearer to or farther from you—as well as tilt them or rotate them about an axis. The image appears on a screen similar to an old-fashioned black-and-white TV screen, except it has a greenish tinge. Early SEMs, like the one we were using when we started, also had a moving scan line, so the image scrolled from top to bottom, but it didn't take long to get used to this odd effect. Zooming up in magnification was like floating down into tiny holes or into crevices; backing off was like being on a rocket ship leaving earth. And, unlike light microscopes, which tend to flatten everything, SEM images look three-dimensional, with shadows, depth, and movement, as you "walk" over the surface of this new, tiny world you can see for the first time.

And, of course, the SEM gave us a whole new way to analyze ancient diets. The hyrax teeth, for example, showed dramatic differences from the actions of abrasive food particles on the components from which the teeth are made. The teeth of Bruce's hyraxes, the browsers, were so finely polished by leaves that the individual prisms of enamel (the white substance on the outside of teeth) could be seen. The other component of teeth is dentin, the slightly yellowish material that, in humans, is exposed only when teeth become extremely worn. However, in hyraxes and many other animals, both dentin and enamel are normally exposed on the chewing surfaces of the teeth. Looking at the replicas, we could see the dentin's fine structure clearly too, down to the tiny dentinal tubules that in life house the microscopic arms of the cells that secrete dentin. In contrast, teeth from the grazing rock hyraxes were scratched and scored with hundreds of microscopic grooves. The wear was so heavy that the microscopic structure was completely obscured, whether you focused on the enamel or the dentin. Even during the dry season, when rock hyraxes browse a lot (more like Bruce's hyraxes), this pattern of scratching persists, although the fine leaf polish was also apparent.

In determining what caused the scratches, Enrico was invaluable, for he had also saved vials full of hyrax dung. We knew that grass contained tiny hard particles called grass opaline or phytoliths; it is the grass's main defense against being eaten. When we broke down and dissolved the rock hyrax dung pellets, their main component proved to be grass phytoliths—material certainly hard enough to score the teeth. The leaves of trees and bushes don't contain opaline phytoliths, so eating leaves couldn't make scratches.

We showed that dental microwear could provide valuable clues to ancient diets, even to the point of revealing seasonal variations in diet. The next step was to examine a whole range of animals with more varied diets. Pinpointing the differences in dental microwear and establishing the credibility of the technique took years. None of my colleagues was used to looking at SEM images, much less understanding the microwear I was showing them. But it wasn't long before I had a preliminary answer, satisfying to me, to some of the questions about early hominid diet. My SEM work on *Australopithecus boisei* and small *H. habilis* specimens like 1813 showed a clear dietary difference. They both looked generally frugivorous (fruit-eating), even if graciles seemed to have eaten softer fruit, while the *boisei* teeth were pocked and scored from biting down on hard seeds and fibrous pods. The best dietary analogue for these East African robust australopithecines I could find among modern mammals was something like the chimpanzee or the orangutan. I looked at a few specimens of *Homo erectus* and found extremely harsh wear: the teeth were battered like those of a bone-crushing, meat-eating hyena.

I didn't have any specimens of gracile *africanuses* in Nairobi to study, but Fred Grine, then a student completing his Ph.D. thesis in South Africa, took up the challenge. Fred believes there are certainly two species in South Africa; in fact, he believes robusts are different enough to be accorded their own genus, so he calls them by Broom's old name, *Paranthropus robustus,* rather than *Australopithecus robustus.* He confirmed John Robinson's prediction that there were marked dietary differences between *Paranthropus* and the gracile *Australopithecus;* the former showed heavier wear from more abrasive particles. Both were certainly plant eaters, though not of the same type of plants, and neither was a pure meat eater. However, recent analyses of the chemical composition of australopithecine bones by Andrew Sillen of the University of Cape Town suggest that robusts may have eaten more meat than graciles.

None of this, however, afforded more than a glimpse of the missing link, *Homo erectus,* because I never pursued the question of its diet seriously. As I said earlier, I wasn't much interested in the missing link; it seemed too modern and human to me, and I knew (I thought) what humans were like.

Typically, it was Kamoya who ended our complacency. You see, he can't help listening for early hominids; they call him to come find them. And he doesn't much care what it is—*Australo-pithecus* or *Homo,* skull or jaw or long bone—as long as it is a hom. Analyzing it, making sense of it, squeezing it for information about our past is my job; finding it is his. So it was he who found the first—and the worst—*Homo erectus* skeleton anyone had ever seen. It was a nightmare.

8

FOOD FOR THOUGHT

Kamoya first noticed the fragments of skull and teeth, which he knew immediately were hominid, in 1973, though it took some years before we unraveled their meaning. They were terribly broken up and they seemed to be scattered over an area roughly the size of a football field. This was bad enough, but the area was liberally sprinkled with other fossils too: crocodiles, turtles, hippos, antelopes, elephants, giant baboons, and giraffes, all in hundreds of pieces. In all, the team collected enough fossils to fill five museum drawers (wooden boxes about two and a half feet long by one and a half feet wide) to a depth of several inches and left piles of very large or otherwise obviously nonhominid bones at the site. The fossils were good and strong, and each one had to be examined carefully to see if it was part of the hominid.

Kamoya had discovered the first largely complete skeleton of *Homo erectus* ever found, and we should have been celebrating the way we did later at Nariokotome. Theoretically, a skeleton would mean a tremendous expansion of knowledge that would surely accrue from analyzing both the head and the body of one individual, not to mention seeing the parts of the body that were hardly known before his find. In reality, though, we were less than elated because, by Sod's Law of Paleontology, the individual whose skeleton Kamoya had found suffered from a bone disease that obscured the shape of almost every bone except the cranium. We looked for the diseased bone and plucked those pieces out of the forty thousand or so other fossils. The selected pieces became a specimen numbered KNM-ER 1808, which we pronounced "eighteen-oh-eight," as if it were a name. We knew 1808 was *Homo erectus* from the tooth size and from the partial skull Meave and I reconstructed, and the geological framework told us it was approximately 1.7 million years old.

Because of the extent and severity of the bone disease, identifying the broken bits that belonged to the skeleton was easy. Many of the bones of the arms and legs of 1808 were recognizable, but only just, because they were coated with about half an inch of pathological bone. This bone wasn't smooth, shiny, and normal-looking; it was fibrous and coarse textured, what pathologists and anatomists call "woven bone." Woven bone is rapidly deposited, either when an animal has to grow very fast, when fractures are healing, or when some disease is at work. And yet, where the bones were broken into sections—and they were almost all broken—I could see a core of perfectly normal bone inside this coating of corklike pathological bone. The situation was extraordinary and called for unusual measures. I persuaded Richard to let me take some narrow slices of bone and polish them so they could be used as histological thin sections, something I could give to the pathologists at Johns Hopkins in Baltimore, where I was then working, so they could arrive at a diagnosis.

I was thoroughly intrigued by the specimen. The skull and pieces of pelvis told me that 1808 had been female; teeth and other bones told me she had been an adult. What on earth had happened to her? I knew it wasn't simply extensive fractures, for the pathological bone was too evenly and widely distributed and there was no sign of crooked healing. I knew it wasn't rapid growth either, for although woven bone is laid down during growth, this material was patently pathological. I contacted some radiologists and pathologists I knew at Hopkins. Hopkins is a world-class medical institution, a maze of interlinked buildings that house medical specialists, researchers, and modern high-tech labs that huddle uneasily in one of the poorest neighborhoods in Baltimore. The institution's patient population is a mixture of those whose illnesses require the most specialized care and those too burdened by poverty, ignorance, or hopelessness to resist even ordinary ailments. And now I was introducing a new patient, a distinctly foreign lady who happened to be 1.7 million years old.

I organized a meeting to explain the problem to my medical colleagues and to seek their advice. We all sat around in a conference room over lunch, very serious in our white coats, and I described the facts of the case. (We reenacted the occasion later for Richard's television series, *The Making of Mankind*.) The patient, as she might be described in medical style, was an adult female of African ancestry. Heaven knew if she was black or not, although I would guess that she was, because it is a useful, even necessary adaptation for folks living on the equator. I explained to my colleagues, "She suffered from a disease that has affected her skeletal system, primarily the appendicular bones"—the limbs—"with a coat of woven bone that is up to seven millimeters thick in places. None of the skull bones shows pathological bone. She probably died of the disease."

I showed slides so the pathologists and radiologists could appreciate exactly what we had found. "You see here a thin section of the cortical bone surrounding the medullary cavity of the femur," I continued, as I showed an image of the normal tissue of the interior part of the femoral shaft. "The Haversian system and the osteocytic lacunae are exquisitely preserved," I said, pointing to the bone's vascular network and to the tiny oval cavities where living bone cells had once resided within the tissue. "You can see the beautiful canaliculi"—tiny canals leading outward from each lacuna, through which the bone cells would have extended thin processes, sort of like an amoeba's pseudopods, for communication with other cells—"and the regular, concentric arrangement of layers of normal bone tissue." In fact, the slide looked rather like a picture of hairy raisins embedded in the concentric rings of a cake with circular, not horizontal, layers.

I changed the slide. "Now we've moved laterally, toward the surface of the shaft. The bone is still normal, laid down in regular layers with normal osteocytic lacunae." I moved my pointer to the top of the screen. "But here, something has gone wrong." The regular layers of bone stopped suddenly with one dense layer; beyond it lay wildly disordered tissue. There were partial layers, disrupted layers, or no layers at all within the bony tissue. The osteocytic lacunae were inflated, puffed up into distorted shapes, chaotically arranged. In slide after slide, each moving farther toward the surface of the femur, I showed them the remarkable, diseased tissue.

I asked for their opinion, and the wrangling over the differential diagnosis began. The doctors had a wonderful time pinning down their oldest patient's problem. We decided to write a comprehensive list of all the possibilities, and then to evaluate them. The physicians considered a wide range of diseases that could produce abnormal calcification or bony growth. For each they asked for additional information or symptoms: Was there any pathological bone here, or there? Were there signs of periodontal (gum) disease? What did the X rays look like?

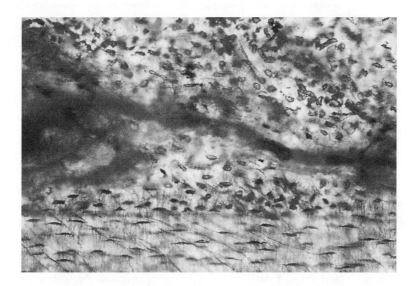

This thin section of part of one of 1808's limb bones, highly magnified, shows the progress of the disease that afflicted this *Homo erectus* female. The section is oriented so that the tissue at the bottom of the photograph lies closer to the central marrow cavity of the bone; the tissue at the top is closer to the outside of the bone. The normal tissue at the bottom was formed earlier in life in regular, well-organized layers. Within these layers are small elongated openings (stained darker and oriented roughly horizontally in this image); these are osteocytic lacunae, minute caves each of which once housed one of the bone cells that created the bony tissue. As each cell surrounded itself with bone tissue, it kept open tiny tunnels, or canaliculi, that communicate with the neighboring cells. These canaliculi look like fine lines that radiate vertically outward from each lacuna. At the top of the photograph, pathological bone tissue laid down near the end of 1808's life is visible. The osteocytic lacunae are bloated and irregular in shape, and even the canaliculi look distorted. In contrast to the tissue shown at the bottom of the image, there are no regular layers of bone. The overall impression is of wildly disordered tissue. This pathological bone coated most of the surfaces of the long bones of 1808.

Slowly a consensus developed. There were two probable options. One was that we were dealing with a disease of unknown origin that no longer existed. "Well, if that's the case," one doctor remarked, "we might as well all go home." The other likely diagnosis set my thoughts off like a Guy Fawkes Day firecracker.

The disease the doctors pinpointed was hypervitaminosis A, an excess of vitamin A. This is a rare disease today, turning up mostly among food faddists, who think if a little is good, more is better (it is easy to poison yourself with vitamin A tablets). But hypervitaminosis A classically occurred among polar explorers, particularly in the early days.

Eskimos and other Arctic peoples have a spate of stories and beliefs that center around the idea that carnivore livers are poisonous, yet it took European explorers decades to realize that these myths and legends were based on hard realities. No matter what its diet, every animal ingests vitamin A throughout its life and stores the excess in its liver, where it is never broken down or detoxified. Carnivores, like dogs, leopard seals, polar bears, or killer whales, eat other animals, including their livers. Because a carnivore eats so many livers, its liver becomes a veritable warehouse of vitamin A. Polar explorers, like the Australian Sir Douglas Mawson, have had an unfortunate habit of getting stranded in habitats where there is little to eat except carnivores.

The Home of the Blizzard, Mawson's account of his epic journey on foot across two thousand miles of Antarctica, is a classic adventure tale that gives grim insights into hypervitaminosis A. Mawson was the leader of an expedition to the South Pole that left Hobart, Australia, late in 1911. From their base camp at Commonwealth Bay in Antarctica, they explored first locally and then more widely. On November 11, 1912, he set out with two men (Belgrave "Cherub" Ninnis and Xavier Mertz), three sledges, and teams of huskies to explore and map hundreds of miles of terrain, relying on provisions they carried with them and some previously deposited caches. On the return journey to the base camp, they realized they were in trouble: traveling too slowly with too little food. They abandoned one damaged sledge and all unnecessary items, loading up the other two. Most of the scientific equipment went on one sledge and the survival gear on another, pulled by the strongest team of dogs. Wary of hidden crevasses, they deliberately sent the scientific sledge first, on the premise that, if it broke through a fragile snow bridge, only the more expendable items would be lost.

On December 13, Ninnis and a full team of dogs plunged to their deaths down an ice crevasse. Tragically, Ninnis had been driving the second sledge, which was gone too, taking with it their tent, spade and pickax, most of their food, all of the dogs' food, their mugs, plates, and spoons, and Mertz's waterproof pants and helmet. Apparently the first sledge had weakened the snow bridge, so the second plunged straight through. Mawson and Mertz were left 320 miles from the base camp with their scientific equipment and rations for two men for ten days. They had already been in extreme circumstances before the tragedy; now it seemed hopeless. They trudged on, killing and eating the sled dogs (and feeding parts of the carcasses to the other huskies) as they went. Soon, suffering from exhaustion, malnutrition, and severe cold, they found the dog livers more palatable than the stringy, tough, musky-tasting muscles, for at least the livers were soft. They didn't know that the livers bore the deadly doses of vitamin A that quickly eroded the explorers' health. They were plagued with dizziness, stumbling, nausea, and stomach cramps. As the poisoning progressed, their hair fell out in handfuls and their skin first cracked and then sloughed off in strips in their boots, hats, and mittens. Their joints ached horribly. Delirium, irrationality, and dementia followed. Because they were suffering simultaneously from scurvy, due to lack of vitamin C, their gums bled and their teeth loosened. As Mertz's dental troubles seemed worse, Mawson gave him the larger portions of liver and chewed determinedly on the repellent muscle himself.

After days of difficult traveling, another crevasse claimed another third of their rations. With true, old-fashioned heroic grit, they went on. Soon they were down to two dogs and a sledge that was far too heavy. They discarded more equipment and killed the dogs. Mertz was terribly ill and soon lost interest in food; when he could, or would, not walk any longer, Mawson dragged the sledge, with Mertz on board. Mawson was staggering. Any sort of movement produced terrible pain, for what they both were experiencing was exactly what happened to 1808. The excess of vitamin A they had eaten—Mawson's biographer reckons they ate sixty toxic doses—caused the periosteum, the tough, fibrous tissue that encases each bone, to rip free from the bone with each pull of a muscle. (The muscles are anchored on bones through the periosteum.) Between the periosteum and bone, torn-apart blood vessels spilled their contents, forcing further separation of the tissues. In the case of 1808, the blood formed huge clots, which ossified—turned to woven bone—before she died.

Mawson was almost as severely affected as Mertz when he stopped pulling and set up a camp to try to nurse Mertz back to health, but Mertz no longer wanted to eat and died one hundred miles from the base camp hut, about three weeks after Ninnis's death and the loss of their supplies. Mertz probably had huge ossified clots on his bones like those on 1808. Mawson buried Mertz in the snow, laboriously cut the sledge in half with his penknife, and blundered on, aching, ill, starving, and unable to think clearly. On February 1, 1913, Mawson arrived at a cache dug into an ice cave only five and a half miles from the camp. He was within days of running out of his pitiful daily ration of eight ounces of food and reveled in the food he found: biscuits, pemmican (the dried meat and fat that is a staple of polar exploration), powdered milk, cocoa, and, improbably, a pineapple.

The weather trapped him in the cave for seven frustrating days. He was tormented by the knowledge that the ship meant to remove his expedition from the Antarctic should have left January 15. When he stumbled into camp, a tottering wreck who had lost half his body weight, he was greeted by his old mate Frank Bickerton, who cried, "My God! Which one are you?"

Mawson's incredible courage and perseverance, as well as his eloquent testimony to the brutal effects of hypervitaminosis A, helped me to understand what poor 1808 suffered. To have such extensive blood clots, she must have been completely immobilized with pain. Yet, despite her agony, she must have survived her poisoning for weeks or maybe months while those clots ossified. How else could her blood clots have been so ubiquitous; how else could they have turned to the thick coating of pathological bone that started us on this quest?

The implication stared me in the face: *someone else took care of her.* Alone, unable to move, delirious, in pain, 1808 wouldn't have lasted two days in the African bush, much less the length of time her skeleton told us she had lived. Someone else brought her water and probably food; unless 1808 lay terribly close to a water source, that meant her helper had some kind of a receptacle to carry water in. And someone else protected her from hyenas, lions, and jackals on the prowl for a tasty morsel that could not run away. Someone else, I couldn't help thinking, sat with her through the long, dark African nights for no good reason except human concern. So, useless as 1808 was for telling us much about normal *Homo erectus* morphology, she told us something quite unexpected. Her bones are poignant testimony to the beginnings of sociality, of strong ties among individuals that came to exceed the bonding and friendship we see among baboons or chimps or other nonhuman primates.

This evidence of sociality needs to be understood in a wider context. All higher primates are intensely social animals, with the possible exception of orangutans, who are more solitary by nature. For good evolutionary reasons, the strongest tie is that between mother and offspring. A mother who pays close attention to her baby, nurtures it, feeds it well, teaches it, and keeps it safe will raise more offspring to adulthood than those who treat their babies more casually and handle them less skillfully. (Indeed, one of the biggest problems in breeding endangered species in captivity is that, deprived of a normal social group, the females frequently fail to learn appropriate mothering skills and the infants die at unprecedented rates due to accidents or simple neglect.) On many occasions, a mother primate has been observed traveling slowly with a sick or even a dead infant, holding the baby to her with one arm rather than expecting it to cling. She may have difficulty keeping up with the group as it travels to feed or sleep, because carrying the infant hampers her movements, but a mother is reluctant to abandon a baby. Older individuals, if ill or hurt, are usually left to struggle along as best they can. Elephants, like humans, are able to show more compassion for the sick. An injured or ill elephant may be attended by one or more others who stay behind with the invalid when the rest of the group moves off, an unthinkable risk for nonhuman primates, who are acutely vulnerable to carnivores if they are separated from their social group. Similarly, sick human children of foraging groups may be left in a base camp or village, attended by their mothers or other relatives, or may be laboriously carried if the entire group is moving.

Where in this continuum of care and bonding does 1808's experience fall? The answer is: well to the human side. The care someone showed her cannot be explained as an example of ordinary mother-offspring bonding. Since 1808 was an adult female, her own mother may not have still been alive. If she were, she would have been past her prime by the standards of that species, which had a short life span; the period of close mother-offspring bonding would be well past. Lacking all evidence, I don't know who cared for 1808 or if it was one or several individuals. Probably, the protector was a relative of some kind, because social groups among early hominids were likely to have been largely coincident with kinship groups. Whoever stayed in the area until 1808 died undertook a serious commitment. The thickness of ossified blood clot coating on her bones tells us that 1808 survived this excruciating and debilitating disease not for a few days but for weeks or even months. The costs and risks of caring for her went well beyond those routinely faced by nonhuman primates or elephants. The bones of 1808 thus speak of the appearance of a truly extraordinary social bond.

There was another message in these bones that became apparent only when I started asking myself, "How did she contract hypervitaminosis A?" The obvious answer is by eating carnivore liver. A Canadian colleague, Mark Skinner, suggested that the source might have been bee brood—eggs, larvae, and pupae, substances said in the literature to be rich in vitamin A. This was an interesting idea, because many primates, and many human groups, prize insects for their rich supply of protein and fat (unappetizing as eating insects seems to Westerners). But when Skinner ran a nutritional analysis on bee brood to check this hypothesis, he found that the amounts of the vitamin were negligible. Carrots or some other vitamin A–rich vegetable might have caused the disease, except that 1808 would have had to eat something like one hundred pounds of carrots versus a pound or so of carnivore liver to produce a comparable effect—not very credible.

If carnivore liver is by far the most likely substance that poisoned 1808, how could she have gotten one to eat? At Koobi Fora the oldest stone tools associated with animal bones date back to about 1.8 million years; elsewhere in East Africa they occur as early as 2.4 million years. The association of tools and animal bones is not an accidental one either. My wife, Pat, was one of the first to realize that the cutmarks she recognized on fossil bones came from stone tools. Where these are found, they are direct proof that hominids used stone tools to process animal carcasses. But cutmarks don't show how the carcasses were obtained, so they don't prove that the "killer ape" epithet can be fairly applied to *Homo erectus*. It is possible, even likely, that during these early periods of human evolution hominids may have scavenged from carnivores' kills rather than hunted. As you might expect, such cutmarks are most commonly found on antelope bones, antelopes being far less formidable prey than carnivores, whose bones very rarely show cutmarks in this time period. Whether or not carnivores were intentionally selected as prey, hominids put themselves into direct competition with the major African carnivores simply by utilizing carcasses of dead animals. Kill sites are dangerous places; scuffles among predators over the possession of a carcass may well lead to death. Did 1808 kill a carnivore in such an interaction and eat it too? Was knowledge of the poisonous nature of carnivore liver slow to be acquired? It seems highly probable.

By the early 1980s we had several glimpses into the diet of ancient hominids. The microwear studies suggested a largely frugivorous diet for the robust australopithecines from East and South Africa, while South African gracile australopithecines had a vegetarian but less coarse and abrasive diet. The skeleton of 1808 told a story of carnivore consumption and predatory behavior, an idea supported by the extreme gouging and battering that showed up in the microwear of *erectus* teeth. Comparable microwear patterns show up only on the teeth of meat-and-bone-eating carnivores, like hyenas. All of this information indicated that hominids had made an important dietary transition from a more plant-based to a more animal-based diet, a change that must have occurred later than *Australopithecus* and *Paranthropus*. What did it entail?

I did not think more SEM work on the East African specimens would answer this question, since postmortem abrasion or chemical erosion had obscured the microwear on most of the teeth. (This is typical of fossils preserved in open-air localities, whereas those in caves such as those in South Africa are more pristine.) My best hope of understanding the dietary transition lay in determining the fundamental costs of becoming a predator. I decided to use a favorite strategy of mine: looking to other mammals. This comparative approach often provides a useful perspective on broad evolutionary questions. By examining other lineages that have coped (evolutionarily) with similar problems, you sometimes get a different slant on our own evolutionary story. It helps free me from homocentrism. What I was looking for were large-scale evolutionary "rules" that would have to be followed no matter what species was in question. My technician, Linda Perez, spent hours in the library gathering data on several crucial points.

The first challenge was to find other examples of herbivores that had turned carnivore. *Carnivore* in this context meant an animal that relies on a diet based on a substantial proportion of meat, not a member of the taxonomic group known as the Carnivora. *Herbivore* as I was using it here meant an animal that eats mostly plant foods of any sort. If these categories seem imprecise, they are, because the diet and ecology of all species of mammals are so highly variable. Some seemingly dedicated carnivores relish the stomach contents of their prey. How much vegetable food does this contribute to their diet? It is an almost impossible question to answer. At the other end of the spectrum, "strict herbivores," such as a giraffe, routinely eat the large placenta and amniotic sac after giving birth. Giraffes (like deer, antelope, and many other species) have also been observed chewing bones in areas where calcium or phosphorus is rare. Besides, by formulating our categories broadly, we hoped to reveal large-scale differences in adaptation that might be detectable in the fossil record.

We wanted to look at animals that had made the sort of dietary transition I knew must have occurred in human evolution, though I couldn't think of many. One was the giant panda, the captivating black-and-white bear that eats bamboo almost exclusively, while most bears are omnivorous or at least partly carnivorous. Pandas could serve as an example of a species that had undergone the hominid dietary transition in reverse, by moving from carnivory to herbivory. Another example, of a herbivore-to-carnivore switch, was a wonderful little creature known as the grasshopper mouse.

The grasshopper mouse is the common name given to any one of several species in the genus *Onychomys* that inhabit the central and western United States. A grasshopper mouse looks like any ordinary field mouse: big ears, bright eyes, bewhiskered nose, long tail, brown furry back, and white belly. But grasshopper mice are ferocious predators that hunt small lizards, other rodents, shrews, and insects, with grasshoppers a favorite prey. They may hunt in pairs or small groups and put up their noses and howl like miniature wolves. The fossil record of grasshopper mice confirms that they are descended from more herbivorous mice.

We couldn't identify many other herbivores that turned into carnivores, so we decided to do an analysis of the adaptations and characteristics of the two dietary groups in general. Obviously herbivores and carnivores differ in their strategies for procuring food, because the food itself has different properties. Plants are unwary, sessile (they won't run away), and can use only passive defense mechanisms, such as thorns, bad-tasting chemicals, phytoliths (which are inedible silica particles), or habits of growth that make them difficult to eat. Plant foods are generally low in protein, high in fiber, and come in natural units (a leaf, pod, fruit, flower, or root) each of which is of low caloric value. What is more, plant foods are usually heavy in indigestible cellulose, so herbivores have to transform their stomachs or guts into fermentation chambers to break down the cellulose and release the nutrients. Prey animals are altogether different.

They are wary, mobile, and use passive (camouflage) and active (biting, kicking, running) defenses. They may be harder to catch, but each unit of food is a big bag of calories: each carcass is high in protein and low in fiber. In other words, the challenge for a herbivore is to *digest* its food, while the challenge for the carnivore is to *capture* it.

This led us to wonder exactly which adaptations are most useful in catching prey. Linda plunged into the ethological literature; her reading of animal behavior studies uncovered some revealing facts. Two attributes of predators account for much of the variation in hunting success among different species. The first is maximum speed of pursuit, which correlates strongly with the percentage of chases that end in a kill: faster predators have higher success rates. A cheetah is more likely to catch the gazelle it is after than is a slower lion, for example. Paradoxically, predators do not necessarily have faster top speeds than their prey. The paradox can be dismantled by thinking about the difference between maximum speed on average for an entire species and the speed obtained by a particular predator during a particular chase. As a predator, you don't have to be faster than every gazelle, just faster than the one you are chasing. The other factor that comes into play is sociality. The social carnivores—for example, lions, hyenas, and Cape hunting dogs—are more successful than you would predict from the general relationship between predator speed and hunting success, while solitary predators—like cheetahs, leopards, or striped hyenas in Africa—do worse than expected. Cooperating to isolate or corner a prey animal and running in relays are effective strategies for predators. So among the most crucial changes for a herbivore evolving into a predator would be anatomical adaptations for speed and behavioral adaptations for sociality.

Once the food item is obtained, it must be processed. Carnivores are notorious for their large, protruding canine teeth, which are used for grasping and killing struggling prey, and for their sharp, slicing cheek teeth known as carnassials. The typical carnivore jaw joint is a tight one that promotes a scissorlike action. Any predator needs a way to slice up its carcass—a far different task from grinding, mashing, or chopping up vegetable matter, which requires broader teeth, sometimes with alternating ridges of enamel and dentin that work like a kitchen grater. Herbivores typically work their jaws in a rotary motion with a great deal of side-to-side action. Even digestion differs in carnivores and herbivores. Meat is readily digestible and requires a relatively shorter gut, most of which is small intestine. Plant matter must be fermented, either in a stomach adapted to become a rumen (a fermentation chamber for breaking down vegetable foods) or in a very long large intestine.

These generalizations, taken as ecological rules, apply to pandas and grasshopper mice in a general way. Relative to other, more carnivorous bears, pandas have much larger, broader teeth, with crescent-shaped crests somewhat like those on deer teeth, for chopping up bamboo. The panda's gut is proportionately longer than that of other bears, although some of this is due to the panda's large body size. Grasshopper mice are a less satisfactory but illuminating example. They have normal teeth for an omnivorous (neither strictly herbivorous nor carnivorous) mouse; they have no slicing carnassials or stabbing canines, but they have large cutting incisors, like all mice. Apparently these are sufficient for dispatching their prey, and their molars are competent to break the prey into pieces. One team of biologists that studied the digestive tracts of grasshopper mice reported that they were fairly normal in size. The team noticed an unusual feature: all the gastric glands are sequestered in a structure called a fundic pouch, which they interpreted as an adaptation to protect these important glands from sharp fragments of the chitinous shell of grasshoppers and other large insects they eat. I would not expect to see a parallel adaptation among hominids.

If acquiring these adaptations to a carnivorous diet is the cost of becoming a predator, there is also an unexpected payback. There is a strong relationship between an animal's body weight and the time it spends feeding: bigger animals have to eat more. This is no surprise. What is more startling is how much the time spent feeding varies with diet. A graph of daily time spent on feeding activities versus body weight for herbivores rises steeply; the same graph for carnivores slopes more gently. In practical terms, this means that a herbivore must spend a much greater percentage of its time feeding than a carnivore of the same weight. For an animal of about seventy pounds, the size of some of the australopithecines, becoming a predator would mean a shift from six to only two hours per day spent feeding, a gain of four hours or one third of the daylight time on the equator. Most carnivores sleep and socialize this extra time away, and that may well have been true of early hominids. But what if this free time were crucial to the innovation and creativity upon which we hominids pride ourselves? Perhaps you must have a substantial amount of time for socializing or idly manipulating objects before you can develop true language or invent the first tools. The idea haunted me.

Another important aspect in becoming a predator derives from a simple principle known as the Eltonian or trophic pyramid. The biomass of different sorts of creatures at different trophic or dietary levels can be diagramed as a pyramid, with each level smaller than the one underneath it. The second law of thermodynamics is at work in the trophic pyramid. All energy on earth initially comes from sunlight, which is then transformed (at a hefty energetic cost) into organisms. The transformation of sunlight to primary producers, or the plants that make up the first step of the Eltonian pyramid, "costs" about 90 percent of the original energy. The next transformation, from primary producers to primary consumers (herbivores, or the second, smaller step of the pyramid), costs another 90 percent, leaving 1 percent of the original energy. Yet another transformation, from primary consumers to secondary consumers (carnivores, which comprise the third, still smaller step of the pyramid), takes another 90 percent of the remaining energy, leaving 0.1 percent. Where tertiary consumers exist (carnivores that eat other carnivores, the fourth, smallest step), only 0.01 percent of the energy is available, which is why we don't eat lionburgers. The inescapable nature of the trophic pyramid explains the common observation that there are always fewer predators than prey. There have to be, or else the predators would quickly run out of food. This explains why predators always have much larger home ranges than do their prey. Similarly, there are always fewer herbivores than plants. The real issue is biomass (weight of organisms per unit area), of course, not number of individuals.

For a herbivore who is evolving into a predator, there is a serious density dilemma. Unless that species already exists at unusually low densities for a herbivore, it must lower its density in a dramatic fashion. Two obvious means to accomplish this end are eliminating most of its total population (fewer individuals over the same area) or spreading that population over a much larger geographic range (the same number of individuals over more area). A third solution would be to diminish the body size of the species markedly (the same number of individuals, but a smaller biomass over the same area). While a drop in population numbers would have been difficult to prove in the fossil record, the other two solutions to the density dilemma—expand your geographic range or diminish your size—ought to be readily visible.

Apparently home-range size is easier to adjust in response to dietary changes than body size, for both pandas and grasshopper mice have done just this. Herbivorous pandas have higher population densities and smaller home ranges than do more carnivorous bears. And grasshopper mice, which have undergone the opposite dietary transition from herbivory to carnivory (like hominids), have kept their body size stable, while increasing home range size and thus decreasing population densities.

All of these points boiled down into a few predictions about what would happen when a herbivore evolved into a carnivore, predictions that would change my perception of the hominid fossil record. First, hominids that turned carnivorous would show adaptations either for greater speed or for social behavior, or both, unless they already possessed these adaptations prior to the transition. I felt I could check this one off: 1808 certainly gave evidence of sociality at 1.7 million years. And, although I did not really comprehend it at the time, 1808 was long-legged and slim, with longer forearms than upper arms and longer legs below the knee: adaptations for running. Second, hominids would have to develop means of slicing up animal foods, either through dental adaptations (which I knew did not exist) or through technology. Stone tools are the obvious adaptation, but they were first found at 2.4 million years, preceding *Homo erectus* by about 600,000 years or so. Third, carnivorous hominids ought to have smaller guts overall, with longer or more elaborate small intestines. At the time, I couldn't see how we would ever discover this from the fossil record. Fourth, newly predatory hominids would gain substantial amounts of free time that might reveal itself in technological innovations or improved means of communication. Again, this point seemed to lead me either to 2.4 million years ago, the time of the invention of stone tools, or to 1.9 million years ago, the time of the first appearance of *Homo habilis* (or big-brained things with australopithecine faces like 1470 and its conspecifics, whatever you called them). Since the skull of 1470 showed the recognizable imprint of a Broca's area in its brain, anthropologists generally assumed that language had also developed by then. Fifth, hominids would have to resolve the density dilemma by shrinking their body size, dropping their population numbers, or expanding their geographic range, and the latter seemed most likely. As expected, I could see no trend toward a reduction in body size throughout human evolution, at least not until the invention of agriculture by modern humans. But territorial expansion certainly occurred, most noticeably when *Homo erectus* strode out of Africa to invade the rest of the Old World.

There were two problematic elements of this analysis that made me set it aside for several years. The first was that the predictions generated by these cross-species comparisons were tantalizing, but they seemed to produce a jumble of incongruent answers. One prediction pointed to the invention of stone tools at 2.4 million years; another suggested that the dietary transition occurred with *Homo habilis* or whatever 1470 was; still another was inconclusive; and the last two pointed to *Homo erectus* at about 1.9 million years. Even the clearest indication, the massive expansion of the range of *Homo erectus*, which might be read as a response to increased carnivory, left me at an odd impasse. At the time I was first conducting this analysis, the oldest non-African *Homo erectus* site was thought to be only about 750,000 years old (although Acheulian tools outside of Africa were known from 'Ubeidiya, in Israel, which is certainly more than 1 million years old). If this geographic expansion were driven by a dietary shift, then it should have happened shortly after *erectus* first appeared nearly 2 million years ago. It made no sense that there was a time lag of roughly 1 million years between the appearance of *erectus* and its presence outside of Africa. The only explanation I could think of was that the dietary transition (from herbivory to carnivory) and the species transition (from *Homo habilis* to *Homo erectus*) were uncoupled, and that made little sense too. I did not understand how or why a species would hang around for a million years and then undergo such a dramatic change in ecology without any perceptible physical changes.

Because I could not make a coherent picture out of the evidence, I let the question of *erectus* and its adaptations drift unattended. The missing link was still missing, even though we had quite a few skulls and lots of mandibles from China, Java, and Africa; there were even drawers full of postcranial bones and a partial skeleton, albeit a diseased one. But although I did not appreciate it, the problem was that none of these fossils was good enough to tell us what we wanted to know about *Homo erectus*. It would take WT 15000, the Nariokotome boy, to show us how many clues we had overlooked. It wasn't until then that all these hints became powerful and revealing. And, satisfyingly, in 1991, once I had really begun to understand the boy, I saw a new *Homo erectus* mandible from Russian Georgia that explained one of the last remaining puzzlements. But I'm getting ahead of my story.

9

VITAL STATISTICS ON
A DEAD BOY

It's really all Kamoya's fault that I started thinking about the *Homo erectus* again instead of australopithecines. If he had not walked across the river on a rest day to look in the "wrong" place, I would never have been taken hostage by the missing link. The sheer physical presence of all those good bones from the Nariokotome boy, sitting on the table in front of me, was enough to prod me into action. It was time to pull together all the information about him and about his species that my predecessors like Dubois, Black, and Weidenreich had gleaned and to augment it with what I could find out now, using modern techniques. Within a few months, the boy had commandeered most of my energy.

When *Homo erectus* was first discovered by Dubois, his challenge was to recognize the species and to grapple with the mere fact of human evolution. As evolutionary theory was more widely accepted, the issue for Black and Weidenreich became *how* to analyze the fossils: which measurements to take, what features to note, and, eventually, which statistics to apply in comparing two forms or two populations. Attention had shifted from Asia to Africa; Dart, Broom, and the Leakeys had filled in gap after gap in the fossil record, with specimens that gave rise to innovative and exciting theories. By the time Kamoya found the boy, I could draw on a long history of scientific studies and hypotheses, most based on isolated skulls, or a few postcranial bones. Few paleoanthropologists before me had the luxury of a nearly complete skeleton, so my challenge was to extract the full meaning of this superb specimen that had been Kamoya's birthday gift to me. I decided to apply every technique and trick of modern science, every approach, however novel, that I could dream up.

Unfortunately, Richard was not going to be working with me as he always had before, contrary to both our expectations when we started. In 1988, when we finished excavating at Nariokotome, the poaching of elephants in Kenya had increased so alarmingly that Richard felt compelled to speak out publicly. Using his position as chairman of the East African Wild Life Society, he began criticizing the government's management of the game parks and wildlife—a risky endeavor in a Third World country—and, being Richard, he managed to attract international publicity for his outspoken views. Early in 1989, in a bold move, President Daniel arap Moi of Kenya abruptly declared that Richard would now be the new director of the Department of Wildlife Conservation and Management, responsible for fighting poachers. Richard learned of this upheaval in the same way everyone else in Kenya did: by listening to the news on the radio.

After some months of uncertainty, the old bureaucratic structure was dismantled, and Richard was put in charge of a new organization called Kenya Wildlife Service. He handpicked the 4,500 men in his charge, which allowed him to replace those whom he suspected of ignoring or even profiting from the poaching, even if he lacked sufficient evidence to prosecute them legally. He embarked on a massive program to raise funds for new training and modern equipment for rangers, improving their working conditions and park facilities. This infusion of determination and money created a new esprit de corps and raised morale among the game rangers, who could now track and combat poachers on an equal footing. Moi even passed a new law that allowed rangers to shoot poachers on sight. The net result of these changes was that elephant poaching diminished with amazing rapidity. But, from the perspective of paleoanthropology, the pressing need to save Kenya's wildlife deprived the field of one of its ablest spokesmen and most successful expedition leaders. Saving the wildlife was a massive task that took all of Richard's skill and energy. Sadly, in 1994, he was forced to resign under intense political pressure, although he had accomplished much. Still, I don't think he will ever come back to paleoanthropology as a primary occupation; his attention has turned to more global matters.

I was not going to analyze the skeleton alone, so my intention from the outset was to search out the smartest young scientists I could find—*smart* because I wanted new ideas, *young* because they would be hungrier for the opportunity, less set in their ways, more creative. As the analysis proceeded and questions arose, I slowly put together a team by calling on young colleagues with all sorts of diverse expertise. It was the chance of a lifetime.

When the excavation at Nariokotome ended in 1988, much of the slogging, routine work had been completed. Joseph Mutaba and Christopher Kiarie did most of the cleaning of the specimens, using air scribes and dental picks under binocular microscopes, while Meave and Emma Mbua, curator of hominids,

1. Kamoya Kimeu trained and selected each member of the hominid gang, our fossil hunters. He recognized the first scrap of skull of the Nariokotome boy on a small hill near camp.

2. In 1984—early days— Kamoya and I excavate near the thorn tree where the rest of the Nariokotome boy's skull was found.

3. Meave Leakey and I piece together the boy's skull.

8. This historic photograph shows the first *Homo erectus* skullcap, found by Eugène Dubois in Java in 1891. He called it *Pithecanthropus erectus* after the hypothetical missing link between apes and humans first named by Ernst Haeckel.

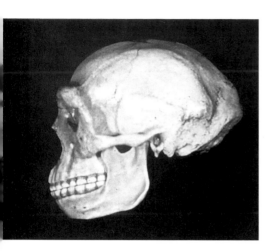

9. This is a composite cast of the skull and mandible of *Homo erectus* from Zhoukoudian, China. The original specimens were lost during World War II and have never been recovered.

10. This photograph is from a mysterious woman who says she has the missing Zhoukoudian bones. The indistinct rounded object in the upper-right-hand corner *may* represent one of the missing skulls, but many of the bones do not match any excavated from Zhoukoudian before the war.

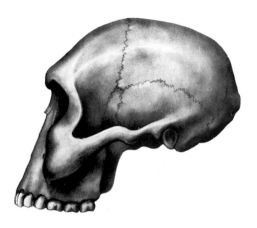

11. Richard Leakey (*below*) holds the face of KNM-ER 1470 vertically and close to the braincase as he discusses the new find with Michael Day. Once we established a solid bony connection between the face and the braincase, the skull's shape was very different (*above*), with a long, sloping face that emphasizes the australopithecine affinities of 1470.

12. Comparing KNM-ER 3733 *(above, left)*, the first *Homo erectus* skull from Kenya, with the more fragmentary *Telanthropus* specimen from South Africa *(above, right)* convinced me that they were the same species.

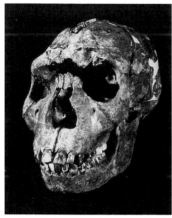

13. This three-quarter portrait of the Nariokotome boy's skull shows his broad, central incisors and his browridge, which would have become bigger if he had lived to adulthood. He was probably about eleven years old in human terms at the time of death.

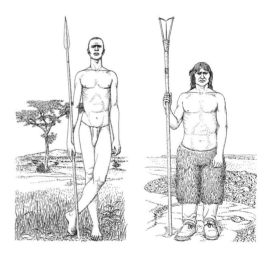

14. The differences in body shape and proportions between a Nilotic tribesman *(left)*, who lives on the equator, and an Eskimo *(right)*, who lives in arctic conditions, reflect their respective needs to dissipate and conserve heat. The boy's elongated proportions show his hyper-tropical adaptation.

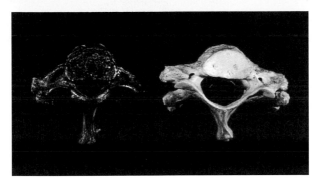

15. The spinal cord passes through the bony canal formed by the vertebrae. The area for the spinal canal in the boy's neck vertebrae *(left)* is about half the size of a modern human's *(right)*. His spinal cord lacked the typical human expansion in the area that commands the muscles for fine control of breathing and speech.

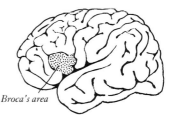

Broca's area

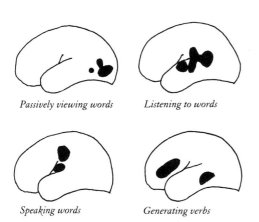

Passively viewing words

Listening to words

Speaking words

Generating verbs

16. I initially believed that the boy had language because I could see the impression on the inside of the left temporal region of the skull by Broca's area, long believed to be a speech-generating center in the brain. This drawing shows the left side of the brain of a person who is looking left.

17. Modern studies using PET (positron emission tomography) scans of the brains of normal humans, with active areas darkened, show that the classic Broca's area does not always function during linguistic tasks, greatly weakening the evidence for speech in the boy.

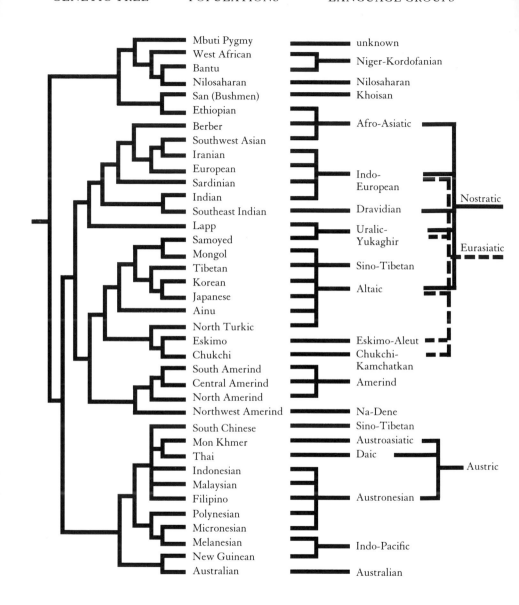

| GENETIC TREE | POPULATIONS | LANGUAGE GROUPS |

Populations and language groups:

Mbuti Pygmy — unknown
West African
Bantu — Niger-Kordofanian
Nilosaharan — Nilosaharan
San (Bushmen) — Khoisan
Ethiopian
Berber — Afro-Asiatic
Southwest Asian
Iranian
European — Indo-European
Sardinian
Indian
Southeast Indian — Dravidian
Lapp — Uralic-Yukaghir
Samoyed
Mongol — Sino-Tibetan
Tibetan
Korean — Altaic
Japanese
Ainu
North Turkic
Eskimo — Eskimo-Aleut
Chukchi — Chukchi-Kamchatkan
South Amerind
Central Amerind — Amerind
North Amerind
Northwest Amerind — Na-Dene
South Chinese — Sino-Tibetan
Mon Khmer — Austroasiatic
Thai — Daic
Indonesian
Malaysian
Filipino — Austronesian
Polynesian
Micronesian
Melanesian — Indo-Pacific
New Guinean
Australian — Australian

Nostratic
Eurasiatic
Austric

18. Luigi Cavalli-Sforza showed that there is a stunning concordance between trees of genetic resemblance of different human populations *(left)* and trees of linguistic similarity of those same peoples *(right)*. If true language arose very early, the language tree ought to be much more diversified than the genetic data.

undertook to number and catalogue the hundreds of fragments. In the meantime, I worked at measuring and writing formal descriptions of all the pieces. A complete human skeleton has 206 bones, counting lefts and rights of the same bone, but if you count all the paired bones only once—on the "seen one humerus you've seen 'em both" principle—then there are 120 bones in what might be called a half skeleton (meaning one arm, one leg, and all the bones of the head and torso). We had 67 bones, or 33 percent of the whole skeleton and 40 percent of a half skeleton. This figure struck me as odd, because it is often repeated that the *Australopithecus afarensis* skeleton known as Lucy is about 40 percent complete—and yet we had more of the boy's bones than of Lucy's. Calculating her completeness in the same way as I had for the boy, and allowing even a fragment of a bone to be counted as complete, Lucy is only 20 percent of a whole skeleton or 28 percent of a half skeleton. Puzzled, I asked Don Johanson how he had arrived at his figure for Lucy. His answer: when Lucy's completeness was calculated, he discounted the 106 bones of the hands and feet, perhaps because they are so rarely found; by the same procedure, the boy is 66 percent complete. But, of course, Lucy's bones were not the main point; the boy's were.

When the excavation was over and I knew that no new bones would be found, I decided that the place to start the analysis was with the vital statistics of the dead boy: sex, age, weight, height, and brain size. I had a rough feeling for most of these points— it was impossible not to speculate about them as I cleaned and described the bones—but now the analytical team had to flesh out the skeleton.

I knew that the boy was immature, because we found diaphyses, or shafts, of the long bones in the excavation as well as the unfused epiphyses, or articular ends, that once fitted onto them. From the ruggedness of the skull and the already large browridge, it was equally obvious that he was male, since a female would have been more lightly built. Detailed measurements of the pelvis, the portion of the skeleton that is the most reliable indicator of sex, confirmed it: he was certainly male and also markedly slim-hipped. Sex was the first point to establish, as it so often is with any new acquaintance with whom you intend to spend a lot of time. Any parent knows that boys lag behind girls in maturation, both physically and emotionally, but this is not just a human trait: it is true of many mammals. Once I was sure our specimen was male, I could ask how old he was and expect a meaningful answer.

The first crude approximation was derived from the teeth, for his second molars were newly erupted, an event that occurs at about twelve years of age in boys today. The second approximation came from the bones themselves. Since not one bone had all its epiphyses fused, the boy was certainly younger than sixteen or seventeen years old, in human terms. That was an upper limit on his age. The lower limit could be established by looking at the only bone that showed even the beginnings of epiphyseal fusion: the humerus. In humans, the distal, or elbow, end of the humerus is the first region to show epiphyseal fusion. There are actually four separate epiphyses and, on the boy, one of these had definitely fused and another was on the way. This process begins between the ages of eleven or twelve and fifteen, with variations according to the population in question. Because the innominate or hipbone, another site of early fusion, was not fused, we could narrow the range still further. Sometime between the ages of thirteen and fifteen, bony fusion begins to join the three separate elements of the innominate (the ilium, the ischium, and the pubis). This meant he was probably less than fifteen and quite possibly less than thirteen. Skeletally, he seemed to be between the ages of eleven and thirteen *if he were a modern human male.* That was the big "if," because if we knew anything, we knew he was not a modern human male. And humans, male and female, have some extraordinary aspects to their growth and development. Finding out the boy's true age would not be a straightforward task.

I turned for help to B. Holly Smith at the University of Michigan. A petite blonde with a ready laugh, Holly is an expert in age determination and in the different patterns of growth and development in primates. In a field marked by complicated methodology, recalcitrant data, and subtle evidence, she is building a solid reputation through her elegant work and intelligent interpretations. Holly has painstakingly studied hundreds of immature skulls of monkeys, apes, and humans using radiographs and direct observations. These data have permitted her to refine previous methods for determining age at death and have given her a unique perspective on the significance of timing shifts in dental maturation. I knew she had the background to give me the boy's precise age at death and the brains to unravel what his maturational state might imply about the life of *Homo erectus*.

At a meeting of the American Association of Physical Anthropologists in 1989, I asked Holly if she wanted to work on the project. She was anxious to collaborate. She had seen a brief television news clip then when we found the specimen, and she had read our preliminary publication in *Nature* and a popular article in *National Geographic*. She also knew there was very little data about *Homo erectus*'s growth pattern at the time. "I wasn't going to pass that opportunity up," Holly said to me later, remembering the occasion. "I didn't fool around. You asked me in early April and I was there in Baltimore on May 12."

I had cleaned and reassembled 15K's bones in Nairobi, and Simon Kasinga, in the casting department, and I had made casts for Holly of all of the teeth, including the ones that had fallen out of the jaw. We also took X rays of the jaws for her. She sat in my lab at Hopkins and pored over this information. "They were really nice, sharp casts," she remembered later, "quite beautiful. I could see every tooth on at least one side. There were nice X rays of the mandible. You could see everything. I sat there and just drew them all, very slowly, to get enough information so I could estimate a development stage for each tooth." The wealth of material—having all of the teeth and most of the skeleton—raised its own peculiar difficulties for Holly. Originally she intended to write a short note, she told me. "I'm still embarrassed that I saw twelve teeth and wrote fifty pages. But everyone who tackled this skeleton found out that every time you had a question, you had to do a huge research project to answer it."

Holly's first problem was to decide which referential model was most appropriate for analyzing the specimen. As Holly put it, "Is he 'like humans,' 'like chimps,' or 'like monkeys'? That's what we needed to find out." Referential models are those based on what we know about a specific living genus or species. Each of the three models (the human, the chimp or ape, and the macaque or monkey) is based on different assumptions about the pattern and timing of growth of the boy, assigning an age based on which particular events would have occurred during his life history. She decided to start with a truly broad perspective, monkeys to humans, just in case there was something odd about the boy. What difference does the choice of model make? "Humans have a unique and extreme life history among living mammals," Holly told me, "with a long gestation period, a long period of infant and child dependency, and the longest life span of any mammal." Like a striped rubber band, the human life history has the same subdivisions as that of other primates, but particular segments—infancy and childhood—have been stretched out. She also pointed to childhood's abrupt ending with the adolescent growth spurt: a unique feature of our pattern of growth that distinguishes us from all other mammals, including nonhuman primates.

"We're the ones who are odd," said Holly. "In preadolescence, human children are suppressing their growth. We've dipped down or slowed our growth relative to chimps during this period, and then we catch up and rejoin the chimp rate later. Barry Bogin, a researcher at the University of Michigan, has suggested that young mammals signal that they are babies"—through their small size, different coloration, or particular postures, she explained—"in order to make adults treat them differently. The human growth pattern may be a kind of deception that makes preadolescents appear younger than they are; they are certainly oddly small and should be larger. Humans may be 'pretending' to be children a little longer, reflecting our unique human dependence on parents and our longer period of training and learning. It's a way of putting off the troublesome a little longer."

Holly uses various events during maturation as convenient time markers. Building on classic work in the middle of this century by Adolf Schultz, she defines the emergence of the first permanent molar as the end of infancy, the emergence of the second permanent molar (or "twelve-year molar") as the beginning of adolescence, and the emergence of the last tooth, normally the third molar, or "wisdom tooth," as the beginning of physiological maturity or adulthood. In humans, then, infancy ends at about age six; in nonhuman primates, it occurs at the age when infants are usually weaned (three years and four months in chimps and one year, five months in macaques). Humans enter adolescence at about twelve, with the onset of puberty, whereas the same tooth emerges, and the same hormonal changes occur, at about six and a half years in chimps and three years, two to three months in macaques. Physiological adulthood occurs at about age eighteen in humans and coincides with the end of growth in height and the completion of fusion of most parts of the skeleton. The same physiological events also coincide in other primates, except that they occur at about eleven years, five months in chimps and five years, ten months in macaques.

Despite the dramatic differences in the age at which they reach adulthood, humans, chimps, and macaques have startlingly parallel life histories. The steps of the dance of growth are the same, but different species dance to metronomes set at different speeds. "If the timing of the eruption of the last tooth is taken as one hundred percent," Holly told me, "then all three species erupt deciduous, or milk, teeth at five percent and erupt the second molar when they are two-thirds grown. Skeletally, all three species fuse the last of the major epiphyses at about one hundred to one hundred twenty percent." The *relative* age she assigned to the boy might be the same according to any referential model, but the *absolute* age was likely to vary widely.

Yes, but how old was the boy? To answer my question, Holly started with the dental information. The teeth were a far more accurate means of determining age than the bones, because the development and eruption of teeth is such a prolonged and complicated process. Remember, the boy died when the milk canine tooth in his upper jaw was loose and wobbly but not yet falling out: still too painful to pull out. The crown of the permanent canine can be seen clearly, waiting to erupt, but its roots were not yet fully formed.

Holly began by testing all three models (monkey, ape, and human) to see which best fit the boy's state of maturation. She scored the degree of development of each tooth, judging from the casts, photos, and radiographs how close the crown and root were to completion. Inspecting the crowns for wear showed her whether or not the teeth had emerged through the gum yet. She added this to the information about the state of fusion, or lack of fusion, of various skeletal epiphyses and arrived at a first answer. Whether the boy grew like a human, an ape, or a monkey, his relative age was about the same: 65 to 75 percent of the way toward adulthood. He was clearly a young adolescent. While the details of his teeth and skeleton do not match any of the three referent models exactly, Holly calls them "grossly out of synchrony" with the macaque, so that model could be set aside.

Holly's next step was to try a narrower comparison of the boy to chimp and human models, hoping to gain a precise estimate of his age at death and a clearer idea of his pattern of growth. When she compared his exact state of dental eruption—second permanent molars and lower permanent canine erupted, upper canine still emerging—to data on modern humans, she immediately encountered the difficult question of human variability. The problem is that we know too much about human dental eruption. Holly's task would have been easier if there were fewer data. Information that she has compiled from studies on people from all over the world cluster into two groups: people from Europe and Asia erupt their canine teeth later than Native Americans and people from Africa, Australia, and New Guinea. Holly's explanation: "The so-called *twelve-year molars*"—the second molars that usually erupt just prior to the canines—"of Europeans and Asians would be more appropriately called the *eleven-year molars* anywhere in the rest of the world." Not only culture but even physiology can differ in different regions of the world; in this regard, Europeans and Asians are retarded. This dental delay may be partly an effect of the inclusion of more industrial and urban individuals in the Eurasian samples, since food preparation, medical and dental care, and nutritional adequacy may affect growth as well as genetics. As the boy surely wasn't enjoying these advantages, the non-Eurasian samples looked like a more appropriate model for him. By these standards, the boy would have been about 11.1 years old at the time of death. Looking closely at the formation of the various teeth, Holly then revised this estimate downward slightly, to an even 11 years. This age coincided well with the lower end of her estimate based on his skeletal maturation, so we both accepted it for the time being.

Holly then went through the same procedure of calculating 15K's age at death as if he were a chimpanzee. In that case, his teeth suggested an age of about 7 years from dental eruption data, perhaps 7.5 to 8 years based on dental development, and about 7.5 years based on epiphyseal fusion. Which model was more accurate? Here, Holly used a clever means of looking at the data. She constructed a special sort of graph, in which the *X* axis represented years and the *Y* axis represented the development of different teeth. One such graph was based on human data, so that plotting the development of a human individual's teeth would yield a straight vertical line. Another graph represented chimps, with an individual chimp's data yielding a straight line. Then she plotted the boy's data on these graphs; the straightness or crookedness of the line would reveal how humanlike or chimplike *Homo erectus* was.

Neither the upper nor the lower teeth of the boy made a straight line on any graph: the answer was not "human" or "chimp" but "none of the above." The boy differed from humans in the early eruption of the molar teeth; he differed, rather widely, from the chimp or great ape model in the very late eruption of the canine teeth. In other words, he had not yet evolved a human pattern of dental development, but he seemed to be moving toward one and away from the apelike condition that is presumed to be ancestral.

If neither pattern fit the boy well, then neither age estimate (11 years or 7.5 years) was correct. The absolute time scales Holly used to make the prediction needed to be adjusted. But how? Fortunately, in an earlier study Holly had discovered that brain size is a very sensitive predictor of the age at which various events occur during the life history of any primate, and some of those events are the emergence of the permanent teeth. We had the boy's skull, so we knew how big his brain was: 880 cc. Using this approach, she calculated an age of between 9 and 10 years for the boy, which, in a *Homo erectus* life span, would make him, once again, a young adolescent. "If you want to know how old I think he was when he died," Holly told me, "I think he was nine or maybe nine and a half. But I think he grew up on a time scale halfway between that of chimps and humans, so that number is misleading. He was not comparable to a nine-year-old human socially or emotionally; he was comparable to a twelve- or thirteen-year-old," because *Homo erectus* matured more rapidly than humans do.

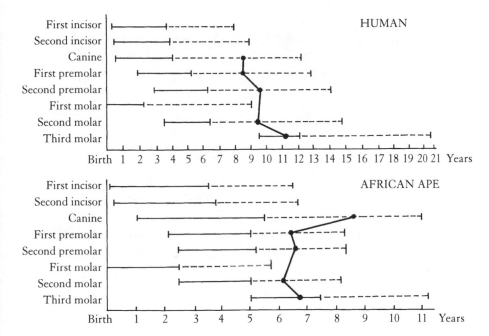

Holly Smith compared the Nariokotome boy's dental development with that of humans and chimps. She plotted the range of ages during which each tooth crown developed (thin, solid line) and each tooth root formed (dotted line); a tooth erupts through the gum during the period of root formation. Then she superimposed information about the boy's dental maturity in bold black circles and lines on both the human and chimp patterns. If the boy's developmental pattern matched either one, his data would form a straight vertical line. Whether she analyzed upper teeth or lower teeth (as here), the crookedness of the boy's line showed Holly that he had a unique pattern of physical development that was probably typical of all *Homo erectuses*.

At this point, two aspects of the project collided. I had been working with Christopher Ruff, an anthropologist who was also at Johns Hopkins, to calculate the boy's stature (height), weight, and body proportions. Since the early days of excavation, I had realized that the boy was tall, because I held his leg bones up to the legs of the Turkana kids who worked for us. Some of the early news reports, including a Tom Brokaw broadcast that Holly remembered seeing, dubbed him "the strapping youth" on account of his height. To calculate a more precise estimate, Chris and I used a well-known method based on the relationship between the length of the femur (thighbone) or tibia (major bone of the lower leg) and stature in modern humans. This technique yielded an estimated stature of about five feet three inches; maybe the boy would have been an inch or so shorter, because his lower-than-human cranial capacity meant that his skull wasn't as domed as a human's would be.

Was five feet three so very tall? It was if he was only nine years old, as Holly assured us he was. In fact, if you make a crude assessment of a child's age by his or her height, he was much too big even for an eleven-year-old. "Even among tall populations," Holly said, "like the Dinka of the Sudan, or the Turkana or Masai of Kenya, boys don't reach five foot three inches tall until about fifteen years of age." Curiously, our boy was nine or ten, maybe eleven, in terms of skeletal maturity, but he was fifteen in terms of height. If he had not yet begun his adolescent growth spurt—if *Homo erectus* had an adolescent growth spurt—he would have been a very tall, lanky man indeed, had he grown up. Chris and I decided to project the boy's adult height. This was not as straightforward a task as it might seem, since we needed to consider which human populations might give us the most accurate estimate. In carrying out this aspect of the analysis, I discovered how intensely focused Chris is. As we worked together, some minor problem would arise, for example, a question of how many individuals had been measured in a particular study. Chris would then leave my office, unwilling to go any further with the work until he had searched through the literature and checked the point exactly. A few days or even a week or two later, he would walk back into my office and pick up the conversation exactly where we had left it off, without preamble. I would look up from whatever other task I was engaged in; with my short attention span and eclectic mind, I rarely had any idea what he was talking about. Chris always seemed surprised that I had forgotten the exact issue that had caused us to suspend the research. Whereas I like to work on sixteen things at once, Chris tends to pursue only one, doggedly and carefully, until he has it just right. Our complementary styles make us good collaborators.

In the end we decided, for reasons I will explain shortly, to use data on the growth trajectories of Nilohamitic tribesmen from equatorial eastern Africa. As a starting point we assumed the boy was age eleven or twelve, Holly's best guess based on human standards. The estimate of adult height we arrived at was a staggering six feet one inch, with a possible height of up to six feet ten inches if he had yet to go into an adolescent growth spurt. As a solid, conservative estimate, allowing for a margin of error in our calculations, we reckoned he would have grown up to be between five feet ten inches and six feet four inches tall. We were stunned—Chris more than I because his previous work on stature showed him that this was clearly an extraordinary height. What had happened to the well-established truism that *Homo erectus* was a short, stocky, muscular species? Even if we had not had the luxury of such a complete skeleton before, we had had more than a few long bones of *Homo erectus*, starting with Eugène Dubois's original femur. If they were so tall, why hadn't anyone noticed before? And, if they weren't, what sort of prehistoric basketball player had we excavated at Nariokotome?

Maybe I was also less surprised than Chris by the boy's height because, after the first field season, when we had recovered his femur and tibia, Richard and I had quickly compared them to the previously known leg bones of *Homo erectus* using the cast collection at the museum in Nairobi. The first thing we noticed, once we got all the casts out on the table, was that the boy wasn't remarkably big, contrary to our impression in the field. He was average-size, in fact, compared to the other specimens. Why hadn't we noticed before that all individuals of *erectus* were so large? The explanation lay in the second thing we noticed: with the exception of Dubois's original *erectus* femur (which some people still think is not an *erectus* but a human), none of the previously discovered *erectus* leg bones was complete. Just for fun, I drew several outlines of the skeleton of *Homo erectus* on a piece of paper and started coloring in the parts that were known at various points in time. Prior to 1981, when we pieced together the diseased skeleton KNM-ER 1808, we had a reasonable representation of the skull and jaw and some good portions of long bones. Not one of the long bones was complete, however. After 1981 we had a clearer picture, except that the bony disease from which 1808 suffered obscured almost everything about its bones. Now that we had good complete limb bones, we could see that our impression of the partial ones had always been wrong.

One reason for our error was that before studying *Homo erectus* we had been studying australopithecines, who were genuinely small in stature. Lucy, one of the earliest specimens and the one (aside from the boy) whose stature can be most reliably reconstructed, was only about three feet six inches as an adult. Granted she was particularly small, but none of the other specimens attested to a large body size. We simply expected *erectus* to be short too and ignored the evidence of the partial bones in front of us. This expectation is reinforced by an experience shared by nearly every child in Europe and America. We have all been taken around historical houses and museums and looked in amazement at the tiny beds, low doors, or short suits of armor. Psychologically, that experience has conditioned us to think of our ancestors as small people, even though the houses, armor, and furniture that created this impression did not belong to human ancestors but to humans. More important, the humans whose accoutrements fill those museums were agriculturalists.

As has been shown by several anthropologists studying remains from different parts of the world, the invention of agri-

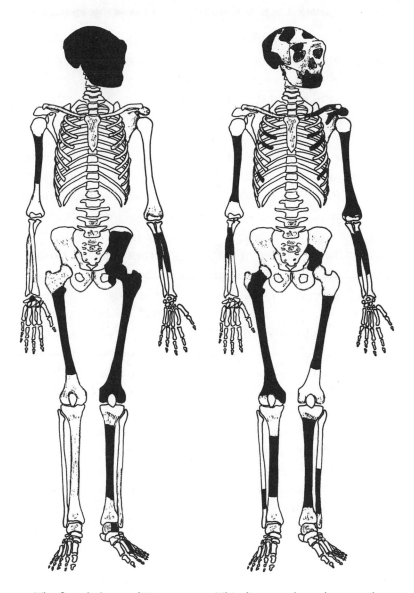

The first skeleton of *Homo erectus*. This diagram shows how much more of the skeleton was represented by KNM-ER 1808 (at right) than all previously known *H. erectus* specimens combined (at left). Unfortunately, the skeleton of 1808 (an adult female individual) was distorted by a bone disease, hypervitaminosis A. The skeleton of the Nariokotome boy was still more complete.

culture was distinctly detrimental to the health of those who first practiced it. Agriculturalists tend to have monotonous diets and recurrent shortages of fresh food; they are also forced to stay in one place and to settle in larger groups, which promotes the transmission of "crowd diseases" like tuberculosis, not to mention creating problems with sanitation and water supplies. Simply put, the way of life and the limited subsistence base of many farmers stunts their growth. Even the industrial revolution at first resulted in a diminution of the health and stature of modern humans. For example, in 1883 the minimum height for recruits to the British army was five feet three inches; in 1900 that standard had to be *lowered* to five feet, so poor was the nutrition of those who wanted to enlist as common soldiers.

Chris and I proved what Richard and I had suspected: the boy was a tall individual who was not unusual for his population. Our stature estimate of five feet three inches as a boy or six feet one inch as an adult was based on information about Nilohamitic tribesmen, people like the Dinka of Sudan who live close to where the boy's fossilized bones were discovered. We had used the same dataset to estimate the boy's body weight at about 106 pounds at the time of death and about 150 pounds if he had grown to adulthood. He was, by anybody's reckoning, a long, skinny kid who would have grown up to be a beanpole of a man. We did similar calculations for five other *Homo erectus* individuals from East Africa whose bones had been found previously. Adding them to the boy's estimate, we found that *Homo erectus* in Africa had an average stature of five feet seven inches (ranging from five feet two inches to six feet one inch) and an average weight of 128 pounds. I calculated that, if these individuals were alive today and if by some bizarre chance they were all males, this *erectus* "population" would rank among the tallest 17 percent of human populations worldwide. But it is far more likely that some of these individuals were females (for that is what their anatomy indicates to me). In that case, the *erectus* population would rank among an even smaller percentage of the world's populations in terms of height. Their size is truly astounding.

We were not grounding our estimates on the undemonstrated assumption that people who live in that region of eastern Africa are somehow more closely related to the ancient hominids who lived there 1.5 million years earlier. After all, 15K was an entirely different species from the modern humans of the southern Sudan or northern Kenya today. And Chris's careful forays through the available data on leg bone length and height had demonstrated to us that sharing African descent was not enough to make a particular population a reliable reference sample. There is, for example, a great deal of information in the literature about the size, shape, and proportions of the bodies of Americans, both male and female. But when Chris compared data on African Americans, Americans of Caucasian ancestry, and African blacks, he found remarkable discrepancies. In terms of the relationship between tibial length and stature, Chris has concluded that the U.S. blacks are as different from African blacks as they are from U.S. whites. Obviously, body shape and proportion are both labile and variable. They seem to respond to environment, or more specifically, ambient temperature and the demands of thermoregulation. Therefore, we chose the reference sample for our stature estimate to incorporate one of the most fascinating things we had discovered about the boy: he showed exactly the same physiological adaptations to climate 1.5 million years ago that some modern humans do.

Like other mammals, humans adapt to the environment in which they live in various physical ways. One of the most obvious is through body proportions, following what is known as Allen's rule, which states that people (and other mammals) adapt to the ambient temperature of their environment through variations in limb length. Hotter climates pose problems of heat dissipation. Where it is dry and hot, natural selection favors those with long, slender limbs that increase the body's surface area and thus help dissipate heat through sweating. (In hot, wet climates, like rain forests, sweating is less effective. People adapted to those environments often have the same body width as those in hot, dry areas but are short, like Pygmies, to keep their body mass low.) Conversely, cold climates make heat retention difficult, conditions that favor individuals with short, stubby limbs and a low ratio of surface area to volume. The classic contrast in anthropological textbooks is side-by-side photographs of an Arctic Eskimo (short arms, legs, fingers, and toes) and an equatorial Dinka (elongated arms, legs, fingers, and toes). Their bodies reflect the inescapable demands of thermoregulation.

The measurements most often used to demonstrate the working of Allen's rule are the brachial index (the ratio of the length of the upper arm to that of the forearm) and the crural index (the ratio of the length of the thigh to that of the lower leg). High values in these ratios mean that an individual is cold-adapted; low values indicate heat adaptations, because it is the more distal, or peripheral, parts of the body that lengthen. Fortunately, we had the right body parts to calculate both indices. We found that, compared to living African (tropical climate) and living European (temperate climate) populations, the boy was extraordinary. He was not simply tropical; he was hypertropical, with ratios below the range of ratios recorded for living Africans.

The antiquity of this bodily adaptation to heat stress told us something else too: the boy was probably running around in hot, open country and *sweating*. Although it is indirect evidence, the boy's body build suggests that he, and all *Homo erectuses,* had lost whatever body fur or hair our more ancient ancestors probably possessed. If he had been hairy or furry, then panting (and avoiding activity during the hottest hours of the day) would have served as his main mechanism for heat loss, as it does for the other animals of the African savanna. Because he had no furry protection from the harsh equatorial sun, the boy was probably also very darkly pigmented. Another fascinating feature reflects another aspect of the boy's adaptation to functioning in such a hot and arid environment: he had a nose, a real nose. Bob Franciscus and Erik Trinkaus of the University of New Mexico documented the fact that *erectus* was the earliest species to have a projecting, human-type nose. In contrast, australopithecines and *Homo habilis* had flat, apelike noses, so that (prior to *erectus*) the nostrils leading to the nasal aperture—the opening of the respiratory tract through the nostrils—were sunken into the surface of the face rather than being part of an external nose. The analysis conducted by Bob and Erik suggested that this new shape of the nose allowed a greater volume of incoming air to be moisturized before it reached the lungs and yet also permitted moisture to be conserved and reclaimed as air was exhaled. Most mammals, presumably including australopithecines, possess effective physiological and behavioral mechanisms for conserving water, such as resting more and moving less when temperatures are high or keeping to the shade whenever possible. You have to ask why *Homo erectus* would need to add another mechanism: What had changed? The answer must be related to activity patterns rather than climate or habitat, since *Homo erectus* lived in many of the same places and with the same animals as *Australopithecus* or *Homo habilis*. Retaining the humidity of the nasal and respiratory mucosa would be an important improvement that would enhance the boy's ability to be active—running, walking hard, digging, carrying things, climbing trees—even during the hottest part of the day. Like his long legs, narrow pelvis, and energetically efficient body build, the boy's nose was yet another signature that he was a member of a species that (like later Englishmen) went out in the midday sun.

Another, more generalized way of adapting to hot climates is through relative body breadth. Chris models body breadth crudely as a cylinder, with hip width providing the diameter of the cylinder and body height determining its length. In the

Eskimo-Dinka comparison, you cannot help seeing that the body shapes of these two groups vary as well as the limb lengths: Eskimos tend to be short and stocky (a short, squat cylinder), Dinka tend to be long and lean (an elongated, narrow cylinder). In previous research Chris had shown that the dimensions of the body cylinder are closely correlated with the mean temperature at which different groups live. The boy was slim-hipped and tall, so it was no surprise to find that, once again, his body shape looked most like that of modern Africans living on the equator. Using a data base developed by Erik Trinkaus, we could even retrodict (predict backward) where such an individual "ought" to live. The answer was that he was adapted to a mean ambient temperature of about 85° to 87° F (or 29.2° to 30.8° C), just about what is recorded for the Lake Turkana region today. It was a satisfying confirmation of our interpretation. If the retrodiction had been for an ambient temperature of, say, 50° F (10° C), we would have had to go back and rethink the relationship between body build and temperature. For fun we also retrodicted the mean temperature of a Neandertal, a specimen of the type of fossil humans who inhabited Europe and the Near East during glacial times (about 100,000 to 35,000 years ago). If the boy was hypertropical, then the Neandertal we used fell at the other extreme: he was hyperarctic. The mean annual temperature for Neandertals came out to be 30° F (−1° C), which today would place them inside the Arctic Circle.

We had ample evidence that, from head to toe, the boy was a thoroughly tropically adapted individual. It seemed impossible that he would compromise these adaptations as he grew up, which meant that, had he lived, the boy would have been likely to follow the same rules of growth as do other people with similar body proportions and tropical adaptations. This was why we used information from Nilohamitic tribesmen in reconstructing the boy's stature from the length of his femur or tibia. It was not a matter of believing they shared a closer *genetic* relationship

with *Homo erectus*; rather, we knew that these modern populations shared a common *environment* with *Homo erectus* to which
they had adapted similarly.

We also found out that the boy was incredibly strong. One of
Chris's areas of expertise is in the application of engineering
beam theory to human bones. Simply put, beam theory demonstrates that the way to make a beam stronger (more resistant to
breaking) is to arrange the substance of the object so that most of
the material is aligned along the axis where the beam is subjected to the greatest stresses. A beam subjected to equal stresses
from all directions ought to be cylindrical, with a circular cross
section. But, for example, a horizontal beam that is most likely
to be bent along a vertical or superior-inferior axis should be
shaped differently to prevent bending. Thus, this beam's cross
section ought to be thicker from top to bottom (on the superior-
inferior axis) than side to side.

Now substitute a long bone like a femur for the beam, change
the beam's orientation from horizontal to vertical, and make the
predominant stress anterior-posterior rather than superior-
inferior. Although a femur or any other long bone is basically
cylindrical in shape, the femur has remodeled to increase its mechanical strength, so that its cross section is elliptical rather than
circular. The placement of the extra bone tissue—the material
that thickens the bone in an anterior-posterior axis—tells you
the axis in which the greatest, habitual stress was produced by
the muscular actions of the owner of the femur. The main difference between beams and bones is that beams, once manufactured, do not change their dimensions, whereas bones respond
to the stresses caused by movements and usage by remodeling
(adding or subtracting bone tissue) throughout an individual's
life. Take away those stresses, or change them, and the bones
will remodel once again.

With Trinkaus and Clark Larsen, two colleagues who are experts on Neandertals, Chris and I surveyed the mechanical

strength of the femurs of a chronological sequence of fossils from *Homo habilis* to *Homo erectus* to archaic *Homo sapiens* (Neandertals) to modern humans. The earlier members of our genus, including the Nariokotome boy, had incredibly robust bones compared to ours. In fact, we could document a steady decline in the robustness of the femur at midshaft (measured as the area of cortical bone in a cross section) through our sequence. This change is not just a downward trend; it is an exponential decline in cortical area (standardized for body weight) over a period from 1.89 million years to today, which implies an exponential decline in activity or mechanical stress on the femur through that period. Even as an adolescent, 15K apparently maintained remarkably high activity levels, for he resembles adults of *Homo erectus* in this regard. As a result, he attained exceptional strength relative to modern humans.

I couldn't help but smile inwardly at the image I was developing of this boy: all long legs, jutting elbows and knees, and probably a klutz—superbly strong but tripping over his own feet. It reminded me just a bit of my own son, Simon, when he was a tall, skinny adolescent, only Simon did not have the boy's literally inhuman strength. It put me in mind of Eugène Dubois, using his shivering son as the model for his statue of *Homo erectus,* even though Dubois had far less fossil evidence on which to conclude that the two were similar. And then I remembered something else: these resemblances were based only on analyses of 15K below the neck. I knew the boy didn't have a skull like Simon's, either in overall shape or in cranial capacity.

At this point, I was pleased that we had learned so many of the words in the boy's physical "vocabulary." His bodily remains spoke to us about his life and what it was like. He was tall and thin and probably black; he lived in a very hot, dry place. Although some aspects of his behavior may have been more apelike than human, he was capable enough to survive to adolescence, to grow very large and become very strong. Perhaps he had lived for only nine years or so, but his species matured so quickly that emotionally he may have been more like a twelve- or thirteen-year-old when he died, face down in that muddy, swampy patch. He lay there, alone, for all those years, waiting for us to find him. We had learned much about his body and the life for which he was adapted, but if we were really going to understand this boy, it was time to focus our attention on the head end.

10

SKELETON KEYS

Because the image of the boy's body was now so clear in my mind, I had started to wonder what he looked like personally, facially; I had only been able to see him from afar until now. Even learning about his size, shape, stature, and body proportions had been important, for this information had radically changed the predominant image of *Homo erectus* to which I and most other anthropologists had once subscribed. But the intellectual quest I was embarked upon was, at its foundation, about understanding this *individual* boy. Because he was in many ways so nearly human, I wanted to be able to look at his face, to recognize him as a person (even if he wasn't a person in the sense that any living human is).

Normally, such reconstructions are done by taking a cast of 15K's skull, overlying it with clay in the form of the appropriate muscles, and drawing a picture of the model of the face. But I wanted to do more than look at 15K's face; I wanted to understand it and see how it would grow and change through life. I turned to Joan Richtsmeier, who worked with me at Hopkins.

Joan is a mathematical wizard who is amazingly adept with computers, one of those people who think in terms of numbers. This sounds as if she is a very dry and serious sort of woman, a bespectacled female computer nerd, but nothing could be further from the truth. Joan combines her intellectual powers with a wickedly irreverent sense of humor. She is the only professor I know who keeps an ordinary-looking Campbell's soup can on her kitchen counter that, when you read the label carefully, apparently contains Sex, Drugs, and Rock and Roll.

Joan's main research interest is in studying and quantifying the changes in skull shape as an individual grows. Much of her work has been collaborative, designed to help the plastic surgeons at Hopkins plan how to help children with craniofacial abnormalities. But she is also fascinated by the ways in which the faces and heads of normal human and nonhuman primates develop and grow. This is a difficult pursuit. Unless there is a specific medical problem, no amount of intellectual curiosity warrants subjecting a normal individual to repeated X rays, nor can an extinct species be regrown in order to measure its skull at various stages. Thus Joan must rely on cross-sectional, not longitudinal, data, by which I mean that she gathers quantitative information about a series of skulls (living or dead) that represent different chronological stages of growth rather than following many single individuals over time. By compiling this quantitative data on overall patterns of growth, she can begin to grapple with how evolutionary changes alter morphology.

The idea is that very small changes at early stages of well-established growth patterns may lead to significantly different shapes in adult forms. In other words, you don't have to radically alter the entire plan of growth to produce a new form; all you have to do is effect an evolutionary change that alters the rate or timing of early developmental events and, by the time the individual grows up, it looks rather different. This concept helps us understand how small evolutionary changes to the genome can create complex changes in shape.

Part of the problem in dealing with growth and development of skulls is one of measurement and comparison. The traditional method for investigating growth and development of the human skull involves taking a predetermined set of measurements on a series of skulls representing different growth stages. To assess the changes over time (through growth), a fixed plane or point of registration is established; a common one is the line connecting the anterior midpoint of the foramen magnum (the large hole through which the spinal cord exits the skull) to the posterior midpoint of the foramen magnum. All other measures are taken from this point or line to various other anatomical points on the face and braincase like the spot on the midline between the orbits, the farthest anterior projection of the browridges, the highest point on the skull vault, and so on. Often ten to twenty such anatomical points can be defined and measured. To display the results, anthropologists then trace the profiles of all of the skulls, aligning them at the registration plane. This creates a classic image that shows the skull-as-onion, with younger and smaller skull profiles contained within the older and larger ones.

Though time-honored, this approach to quantifying skull growth has drawbacks. As Joan says, the problem with this technique is that the registration plane *by definition* cannot grow or alter, and yet it is obvious that all parts of the skull are likely to change and probably shift in relative position as we grow from infancy to adulthood. The second problem is that such a presentation of data is basically two-dimensional, whereas growth obviously occurs in three dimensions. This is why Joan has spent most of her academic career developing complex methods for a registration-free system of shape measurement that integrates three-dimensional information. An ancillary problem has been to develop statistical methods of evaluating the three-dimensional data, so that she can find out whether differences in a particular region are meaningful changes in shape or simply individual-to-individual variation. For some years, Joan has been collaborating in her research with Subash Lele, a professor of Indian descent who is at the Johns Hopkins University School of Hygiene and Public Health. Between the two of them, Joan and Subash have pioneered some new methods of collecting and analyzing data about skull shape.

In Joan's system, points on skulls (called anatomical landmarks) are measured with a three-dimensional digitizer. The skull is fixed in position on the digitizer, and an apparatus that looks rather like a pen tethered by a wire to a computer is carefully pointed at each landmark in turn. The computer records the three-dimensional coordinates of each landmark, slowly creating a detailed dataset of information about the shape of the skull. In this regard, her procedure differs little from the traditional approach, except that the relative placement of landmarks is recorded not as a two-dimensional distance from the registration point or place but as a matrix of three-dimensional distances from each landmark to every other landmark. This complex dataset is analyzed using a method based on engineering principles known as Finite Element Scaling Analysis, or FESA.

In FESA the skull is divided up into smaller geometric units, known as elements, which may be shaped like hexahedrons (six-sided solids), tetrahedrons (four-sided solids), or wedges. The vertex, or corner, of each element is defined as being one of the anatomical landmarks. To study 15K's face, Joan used ten contiguous finite elements (tetrahedrons and wedges) constructed among eighteen landmarks. Joan illustrates this measurement system by showing a picture of the skull with what looks like a toothpick lattice attached to it, only the toothpicks are lines that connect the anatomical landmarks. The length of the line indicates the linear distance separating the two landmarks, and the grid formed by all of the interconnecting lines ultimately describes the shape of the face in three dimensions.

Once Joan had measured the boy's face in this way, she had a mathematical dataset representing the original morphology: the starting point. The task was then to compare the boy's face to two different "target" morphologies. These were the average

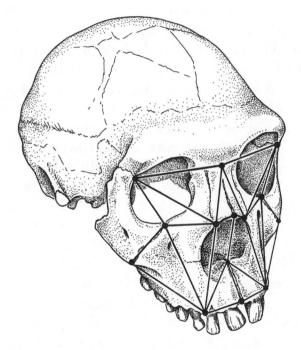

Joan Richtsmeier's mathematically sophisticated analysis of 15K's facial shape started by defining geometric "elements" on the skull, the corners of which are anatomical points. Her research showed, surprisingly, that chimps, humans, and *Homo erectus* look different because they grow along similar trajectories from different starting points. She also found that sexual differences in facial shape appear before adolescence too, giving the boy a distinctly "male" look already.

male and female faces of *Homo erectus*, targets that represented two versions of what 15K's face would have become if he had grown up. Comparing his original face with these adult forms would show Joan exactly how the boy's face would have to grow in order to reach adult shape. For each anatomical landmark, she would be able to say exactly how much it would have to grow and in what directions to proceed from the adolescent to the adult shape. She would be able to say mathematically what a layperson could say intuitively by comparing the boy's skull with that of an adult *erectus*: a little broader here and narrower there, pull this part of the face out, move that down . . .

The results would also give her a glimpse of the rules or pattern of growth that the boy would have followed if he had lived to grow up. Then Joan would be able to ask a question similar to those Holly Smith had wrestled with in looking at dental development: Did the boy belong to a species that grew "like chimps" or "like humans"? To answer those questions, Joan had first to conduct parallel studies of young and adult chimps and young and adult humans, so she would know what their facial growth patterns were.

In all, Joan and I measured five immature human skulls of unknown sex whose dental maturity was comparable to that seen in the boy, four adult human males, four adult human females, four immature chimp skulls at a comparable stage of development, and five adult male and four adult female chimps. We also wanted to study a sample of *Homo erectus* adults, but that was a little harder to come by. We had an adult female with a good face: KNM-ER 3733, the specimen that broke the single species hypothesis. However, there was not one male adult *Homo erectus* skull from anywhere in the world with a largely complete face. Our task was therefore complicated by the need to simulate a male *erectus* face, based on the female one and using the pattern of male-female differences seen in other species. We carried out one simulation using the male-female differences documented in the chimp skulls and another using the pattern of sexual dimorphism observed in the human sample. Each specimen was digitized at least twice, to reduce measurement errors, and the fossil specimens were digitized three times for extra accuracy, which was necessary since we had so few fossils. From the digitized coordinates, we then calculated the measurements of an average juvenile face, an average adult male face, and an average adult female face for both chimp and human. The project involved days of meticulous and painstaking work.

Once we had compiled this huge amount of quantitative information, we could ask how humans and chimps differed in facial growth. As we were documenting them, growth patterns are simply geometric phenomena—the changing arrangements of skeletal structure through time. In both chimps and humans, most of the change and rearrangement of skeletal structure occurred in the upper face, which protruded or grew forward during development, while the lower face elongated along a superior-inferior place. There were some differences between human and chimp growth patterns too. In growing chimps, the upper part of the face widens, especially in the region of the orbits, which widen and tilt slightly upward, a trend not seen in human faces.

Although we expected to find differences, we were more surprised to discover how strikingly similar the human and chimp patterns of growth were to each other, regardless of which sex we were examining. They were more alike than different. Given that fact, we were less surprised to see that *Homo erectus* shared the same overall pattern of growth, whether we considered the chimp- or the human-based simulation of an adult male *erectus* face. This meant that we could only answer "yes" to the question, "Did the boy's face grow like a chimp's or like a human's?" Joan and I puzzled over the significance of this finding. Clearly baby chimps grow up to look like chimps and baby humans grow up to look like humans, and the two are never confused. Eventually we realized the significance of the similarities of growth pattern that we had documented.

First, the similarity in the data among chimp, human, and *erectus* skulls suggests that there is a generalized and perhaps primitive pattern of facial growth that is shared by all adolescent hominoids (a group which includes apes, humans, and their common ancestors). If this is true, then the facial features that are so diagnostic of species—the characteristic arrangement, size, and shape of the nose, cheeks, chin, orbits, and so on—are established very early in life. In fact, it seems possible that these diagnostic differences may even be established prenatally, an idea that ran contrary to all our expectations. Until now, it was axiomatic in biology that embryos, fetuses, and newborns of different species resemble each other more closely than adult forms do. Ernst Haeckel, who created the idea of *Pithecanthropus* and the missing link in the first place, expressed this principle in a catchphrase I remember learning in school: "Ontogeny recapitulates phylogeny." He meant that ontogeny, the process of growth, takes an organism back through the stages of its ancestry, or its phylogeny. Baby monkeys, baby chimps, and baby humans thus ought to look more like each other than the adults, and so most biologists have always believed. But our data now showed that, despite broad superficial resemblances, the distinctive facial template of each species was stamped out very early in development.

All primates share a certain commonality of facial plan, which is part of how we recognize them as primates. Now we suddenly realized that the all-important details, the keys to knowing who is who, must be the result of early and fundamental differences in facial shape. The differences among chimp, human, and *Homo erectus* did not result from evolutionary meddling with the intricate and highly integrated processes of growth and development but from evolutionary changes in the basic face plan. In short, these species do not look different because they grow differently from the same starting point. Instead, they look different because they grow similarly from different starting points.

The second implication was that even subtle sexual characteristics of the face must be determined early in life too. Once again, this finding was contrary to our expectations based on previous knowledge. Osteologists who study the human skeleton have always experienced great difficulty in "sexing"—determining the sex of—immature individuals. The widely accepted explanation for this fact is that many of the sex-specific hormones start to be produced in quantity only at puberty, which is why adolescents are typically moody and temperamental. Before this hormonal change, the skeletons of young humans are sort of asexual, showing no clear differences that can be reliably attributed to sex. Yet our study of the boy's face proved that some of those distinctions already existed. At eleven, the boy's face was more similar to that of an adult male *erectus* than to an adult female *erectus*. We could only conclude either that he had already entered pubescence and started his growth spurt (if he had one) or that the male-female facial differences were so subtle that only a detailed and highly quantitative analysis, such as ours, would reveal them.

All of these findings, put together, suggested that the growth that occurs between childhood and adulthood acts mostly to amplify and fulfill the preexisting male or female and human or chimp or *erectus* morphology. How much earlier than adolescence such sex and species differences might be established we couldn't tell from our study of the boy. He wasn't young enough and we have no other largely complete *erectus* skulls to examine in a comparable way. Until we find such skulls, we won't be able to determine when during growth the species-specific and sex-specific characteristics first appear. Even if we had such skulls in hand, it might take many more years of research to determine. I'll leave that one to Joan.

Growth and development of the boy's face was interesting, but I knew that the size and shape of his brain would be revealing too. Back at the turn of the century, Eugène Dubois had believed that the brain was the key to everything about *Homo erectus,* and I was sure he was right. I enlisted my postdoctoral student, David Begun, to help me estimate the size of the Nariokotome boy's brain and study its shape. When we started the project in 1989, David had just earned his Ph.D. and still looked like a graduate student, with a ponytail, goatee, and gold-rimmed glasses. By the time we finished the project, he had cut his hair, trimmed his beard, and joined the faculty of the University of Toronto, where his fluency in French is a strong asset.

We undertook the simplest task first: estimating the size of the boy's brain. Unlike a human of comparable size, the boy had only 880 cc, not 1,350 cc, with which to control his human-size, long-limbed, and probably gangly body. Of course, to determine this fact, David and I could not measure the actual brain, for soft tissues rarely fossilize. Fortunately, the skull was complete enough that I could make a rubber mold of the inside of the braincase. Because the mold was thin and flexible, I feared it might stretch or fail to retain its shape, so to stabilize it I filled the mold with plaster of Paris while it was still inside the boy's skull. Then I dissolved the glue joins from between the various pieces of 15K's skull until I had a big enough opening to removed the filled, inflated mold safely. The result was a plaster of Paris endocast or inside-of-the-skull/outside-of-the-brain cast. While endocasts don't preserve any of the internal detail of the ancient brain, they do show considerable detail about the brain's surface; the blood vessels that nourish the skull bones, the pattern and distribution of the sulci and gyri that give brains their wrinkled appearance, the size and shape of the major lobes, and various other anatomical features.

The boy's endocast offered no immediate surprises. Everything appeared to be pretty normal for an *erectus* and, when we submerged the plaster endocast in water, we were able to measure its capacity at 880 cc. Next we considered the question of what growth might have done for the boy's brain size. He would not have gotten much brainier as he got older, for (alas) teenage boys have as much brain as they are ever going to have. The same is true of apes at a comparable stage of maturation. Still, if he had lived to adulthood, the Nariokotome boy's cranial capacity would have enlarged somewhat; while the brain does not get bigger after adolescence, the specialized membranes called the meninges that surround the brain thicken and take up extra room inside the skull. This means that the boy's cranial capacity (as opposed to his brain per se) would have enlarged by a bit more than 3 percent, as those of humans do, to about 909 cc. But his brain would remain only about two thirds as big as a modern man's brain—and the boy had a lot of body to control with his brain. By human standards, he had the height of a fifteen-year-old with the brain of a one-year-old, who might normally stand about two feet tall instead of the boy's five feet three inches. It is a staggeringly different view of our ancestors.

We also know that both his brain and his body were pretty average for an African *Homo erectus:* the mean cranial capacity of all known specimens is 910 cc, almost exactly the same as the adult estimate for the boy; the average adult height was five feet seven inches and the average adult weight was 128 pounds. *Homo erectus* specimens from Indonesia and China, some of which are much more recent, have only slightly bigger brains (931 cc and 1,001 cc respectively). Even though brain size *within* modern humans is not correlated with intelligence or mental abilities, it seems likely that brain size *between* species is. The Nariokotome boy probably was not very clever, an assumption that the stone tools of *erectus* (which I have always found rather boring and repetitive) seemed to support. He was an animal as big as a human, who must have looked rather like a human, but he did not have our brains and almost certainly did not act like a human.

We also examined the morphological details preserved on the endocast. There was a healthy Broca's cap, the bump on the left frontal lobe that has been long thought to house a center for the motor control of speech. This region of the brain is named after Paul Broca, a nineteenth-century French anatomist who found that patients with injuries to the brain in this area—from dueling, for example—developed speech difficulties. Their comprehension of words was not altered, but their ability to produce the words was impaired. I pointed out earlier the impression made by Broca's cap to Richard Leakey's daughters when we were first reassembling the boy's skull in camp; although the idea that the boy could talk amused the girls, its presence was not startling to me. Broca's caps have been observed on the endocasts of fossils as old as almost two million years in *Homo habilis,* the species ancestral to *Homo erectus,* so I expected to find it here. Its presence seemed of little consequence but later played an important role in changing my view of *erectus.*

David and I also noticed some asymmetries of the boy's brain, which were more interesting. A former colleague of mine at Harvard Medical School, Marjorie Le May, had documented an unusual correlation between the assymetries of the brain and handedness, using the CT scanner, which takes a series of X-ray "slices" through the head, to look at the brains of living subjects. When the right frontal lobe projects farther forward than the left, the subject is four times more likely to be right-handed; when the right occipital lobe, at the back of the head, projects farther rearward, the subject is twice as likely to be left-handed. Marjorie also showed that the width of the frontal lobes is another good indicator of preferred hand, with right-handers having wider hemispheres in the frontal region about 70 percent of the time. Similarly, left-handed subjects had wider hemispheres in the occipital region. The boy, like most humans, was probably right-handed, for his right frontal lobe projects forward and the right hemisphere is slightly wider in the frontal region. The evidence of the brain was congruent with that from the bones, for the right ulna (one of the bones of the forearm) was slightly longer than the left ulna, as is often the case in right-handers. The clavicles, or collarbones, were similarly assymetrical.

The aspect of the boy's brain that was most intriguing for David and for me was the relationship between brain size and body size. Dubois had pioneered the study of these relationships; he had a complete femur from which to calculate body size, but he had only a skullcap as an indicator of brain size. With the solid estimates of the boy's body size that Chris Ruff and I had calculated and the secure measurements of brain size that David and I had made, we were in a unique position. For the first time, the brain-size-to-body-size ratio of such an ancient hominid could be confidently calculated *for a single individual*. We did not have to rely on body-size estimates from one set of individuals and brain-size estimates from another, lumping them all together by species and hoping the result was reasonably accurate, the pragmatic solution adopted by everyone else. We could even test whether this best-guess double estimate was close to the actual value for the boy.

We used the figures for the boy if he had lived to adulthood in order to make them comparable to previous work. He would have weighed about 150 pounds, or 68 kilograms; his brain size would have been 909 cubic centimeters in volume and 948 grams in weight. Was this relatively brainy, as big brained as a human, or less, more like a chimp? A measure known as the encephalization quotient, or EQ, is the index most often used to compare relative brain size across species. The exact formula for calculating EQ varies slightly, depending upon whether you want to compare your specimen's encephalization to that of all placental mammals, to insectivores (a group taken as representative of the general, ancient mammalian condition), or to anthropoid primates (monkeys and apes). No matter which way David and I calculated EQ, we got a similar pattern of results. The boy's EQ was rather close to the best-guess double estimate for *Homo erectus*—the one based on heads unconnected to bodies—which was a relief. It was also close to the best-guess double estimates for *erectus*'s ancestor, *Homo habilis*. In other words, even though absolute brain size increased markedly between *habilis* and *erectus* (from an average of about 650 cc to about 950 cc), encephalization did not. The larger body size of *erectus* canceled out the increase in brain size. Since the boy coupled a body as large as a modern human—in fact, *larger* than most modern humans—with a brain only two thirds of human size, his EQ was considerably lower than ours.

This was an interesting pattern, reminiscent of Holly's remarks about the human pattern of growth. From her perspective, humans suppress their bodily growth during childhood, making adolescent humans appear younger than they are; then humans, especially males, catch up to the expected rate of growth through the growth spurt, which has a dramatic effect on size. Evolutionarily, brain growth seemed to follow a somewhat similar stop-and-start motif. We could see an abrupt increase in EQ from *Australopithecus* to *Homo habilis*, then an apparent leveling off as both brain and body size increased together as *habilis* evolved into *erectus*, and then another, final abrupt increase (a brain growth spurt?) as *erectus* evolved into *Homo sapiens*.

This is a pattern different from what has been observed among living great apes, where the larger-bodied species have relatively smaller brains, so that gorillas have lower EQs than the much smaller pygmy chimpanzees. But hominids show the opposite pattern: larger-bodied australopithecines have higher EQs than small-bodied ones, and bigger-bodied *Homo sapiens* is also bigger-brained than *erectus*. The apparent exception is that there is no increase in EQ between *habilis* and *erectus*, despite the

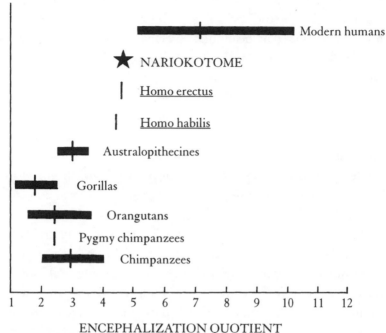

ENCEPHALIZATION QUOTIENT

The encephalization quotient, or EQ, is a ratio that shows how "brainy" an animal is for its body size. The thick horizontal line represents the range of values within a species, while the thin vertical line is the mean or average EQ. Among apes, the larger-bodied species have relatively smaller brains and hence lower EQs. Among hominids, larger bodies mean bigger brains. EQ increases abruptly as body size increases from *Australopithecus* to *Homo habilis*. Encephalization remains about level as both brain and brawn increase during the transition from *H. habilis* to *H. erectus* and then jumps again as *H. erectus* evolves into *Homo sapiens*. Such a profound increase in relative brain size during hominid evolution must have disrupted many fundamental biological processes, producing a conflict between the demands of bipedalism (narrow hips) and those of childbirth (wide hips).

increase in body size. However, given that (by ape standards) a *decrease* in EQ would be expected during this transition, simply maintaining the same EQ probably reflects substantial selection for increasing relative brain size. These selective pressures must have been formidable to override the apelike pattern that our ancestors must once have followed.

As the human lineage acquired large brains relative to their body size, some very fundamental biological processes were disrupted. Humans not only have relatively large brains, they bear relatively and absolutely large-brained babies. Another key characteristic of our species is our unusual mode of locomotion, bipedalism. We are fundamentally two-footed creatures; in fact, one of the proffered definitions of humans is "unfeathered bipeds." Each of these traits would be remarkable in and of itself, but to have evolved both is astounding because the physical and mechanical demands of efficient bipedalism directly oppose those of bearing big-brained babies. Evolution is always a compromise between conflicting demands, but our ancestors tiptoed along a narrow pathway indeed. To move about efficiently, whether on two legs or four, you want to minimize the displacement of your body weight. Staggering—throwing the body weight from side to side—may be easier and more stable for babies or drunks, but it wastes a lot of energy that is expended in propelling the body sideways rather than forward. Four-legged animals generally exhibit very little lateral displacement, and fast, running animals have long, narrow bodies with the left and right legs positioned close together. For example, in horses the side-to-side motions at each joint in the leg are minimized; the fit between the two bones at each joint is so precise that little is possible except fore-and-aft movements. Ideally, bipeds would be built similarly and with their hips close to the center of body weight and each other.

In bipedalism, the trick is to balance the body weight over the hip joint of the supporting or stance leg while the other leg swings forward. If you simply lift your swing leg off the ground, removing one of two supports, you will fall over unless you re-balance your body weight over the stance leg by contracting hip muscles known as abductors. There are two ways to improve the efficiency of the abductors. Either the muscle attachment can be moved farther away from the hip joint—lengthening the lever arm, or the distance between the joint and the muscle force—or the joint can be moved closer to the body's center of gravity (which shortens the so-called load arm). In either case, the same amount of muscle force will produce more movement. The principle is exactly like that used by two children of differing sizes to balance on a teeter-totter; the smaller child has to sit far-ther from the center (lengthening the lever arm) than the larger child.

In terms of adaptations for bipedal locomotion, 15K and other members of his species had used both means of improving efficiency. His narrow hips placed the joints close to the midline, where the body weight is centered; this gave him a short load arm compared to modern humans. Also, his femurs had long necks that connected the shaft to the ball of the ball-and-socket joint of the hip; this arrangement meant he had long lever arms compared to modern humans as well. The net result was that the boy was a more efficient walker and runner than any mod-ern human.

Narrow hips are all very well if you aspire to move effi-ciently—look at the build of Olympic runners and speed skaters—and, as Chris's work on thermoregulation showed, they are an essential component of a body shape that dissipates heat effectively. But there is an overwhelming problem with narrow hips. They hamper the most evolutionarily important endeavor of all: childbirth. Big-brained babies have to pass be-tween the hips, through the bony birth canal, and out into the world. The conflict between locomotion (and cooling) and re-production does not arise in quadrupeds because none have babies with large enough brains to make birth difficult. While the selective pressures for easy births of big-brained babies are enormous, efficient locomotion—for gathering food, avoiding predators, seeking refuge, and meeting potential mates—is critical to survival too. Relative to *Homo erectus* and our earlier ancestors, humans have opted evolutionarily for enlarging the birth canal. This spreads the hips farther apart, lessening locomotor efficiency, but if you can't have any babies to carry your genes into the next generation, you might as well be dead in evolutionary terms anyway. Since *erectuses* had narrower hips, they must also have had smaller-brained babies.

Humans put another twist on the problem of bearing big-brained babies, evolving a tactic that is used by no other mammalian species. Mammals generally follow one of two reproductive strategies. One, known as altriciality, is to give birth to several extremely immature infants at once, creating a litter that must be kept in a nest or den until the youngsters' eyes and ears open and their legs develop sufficient coordination. The alternative plan, called precociality, is to give birth to a single, well-developed infant that can see, hear, run, and jump almost at the moment of birth. Cats with their helpless kittens are a good example of an altricial species, while wildebeest or zebras, with their babies who follow the herd within minutes of birth, exemplify precocial species. When you try to decide which strategy is followed by humans, ambiguities arise. Human babies are obviously not altricial, for they are usually born singly and, from the moment of birth, they are wide-eyed, alert, listening: all their sensory organs are fully operational. In fact, newborns are sensory sinks for information about their new world; they are greedy for taste, touch, smell, sound, and sight. Yet human babies are not precocial, for they are unable to walk or to coordinate their movements in any meaningful way; they cannot even hold on to their parents effectively, as a small chimpanzee or monkey does.

In 1941 Adolf Portmann, a Swiss zoologist, considered this quandary and suggested that humans are secondarily altricial. That is, he sees humans as fundamentally precocial mammals (in terms of our brain development and sensory systems) with an overlay of altriciality that accounts for the incomplete development or retardation of our motor system. Portmann makes some powerful points about the extraordinary differences between humans and other precocial mammals, even apes. From the moment of birth, chimpanzees and gorillas (not to mention antelope or whales) closely resemble adults in limb proportions, body postures, and motor patterns, even if these may be clumsy and labored at first. But newborn humans are incapable of assuming the species-specific upright posture that characterizes their kind. Baby humans simply do not have the physical equipment to stand and walk at birth; the limb proportions must change and the vertebral column must alter until the typical S-curve develops that allows it to act as a support and shock absorber for an upright biped. Amazingly, too, neither upright posture nor bipedality simply appears once the requisite structures are in place; we must *learn* how to sit and then stand and walk upright. No other species must strive to master the art of moving as its adults do. This period of learning is not merely a necessity dictated by the difficulty of bipedal versus quadrupedal locomotion; for example, precocial birds stand bipedally shortly after hatching and often fly well on their first attempt. Astonishingly, although humans are a strongly social species, we are born lacking even the rudiments of the species' means of communication (speech and gesture). These are facts about the human reproductive strategy that are fundamental and unique, that must be explained by any consideration of our evolutionary history.

In the early 1980s Robert Martin started reinvestigating Portmann's work on secondary altriciality and its implications; his insights tell us a lot about the Nariokotome boy. Bob was clever enough to see the importance of the fact that altricial species rarely catch up to their precocial counterparts in terms of adult brain size. This discrepancy occurs because, in all mammals except humans, the brain grows rapidly while the fetus is inside its mother's womb and then declines to a slower rate of growth after birth. The graph of brain weight to body weight of any normal mammal looks the same: there is a straight line (representing a steady rate of growth) until birth, where there is a point of inflection, after which the line slopes less steeply. Altricial species, whose brains are by definition less mature (smaller), simply have no chance to catch up with those of their precocial rivals because of the postnatal slowdown in brain growth.

The underlying mechanism that explains this pattern is fascinating and relies on the functioning of the trophic pyramid that I described earlier. Prior to birth, first the conceptus, then the embryo, and finally the fetus is metabolically part of its mother. Whatever food she eats yields nutrients into her bloodstream, and these cross the placental barrier directly into the bloodstream of the growing offspring. In terms of nutrition, then, the effect of pregnancy is the same as if the mother has simply gained weight. As soon as birth occurs, however, the situation alters radically; the baby is now a separate entity from the mother. Lactation poses a far greater nutritional stress than pregnancy, because the infant is one step up the trophic pyramid from its mother. In effect, the baby is a special sort of parasite that feeds on its mother. The tremendous energy loss that occurs elsewhere in the food chain or trophic pyramid is in effect here too: about 90 percent of the energy the mother takes in is expended in the conversion to her tissues and fluids, including milk. Because the brain is a greedy organ, hugely expensive to grow and maintain, these metabolic demands make it too costly to sustain the rapid fetal rate of brain growth after birth, so brain growth slows.

Humans are the only exception to these generalizations; we have mastered an extraordinary physiological "trick" that keeps our babies' brains growing at the fetal rate for a full year after birth. In effect, gestation in humans is twenty-one months long: nine months in utero and twelve months outside. (When I explain this to an audience, the mothers always nod at this point; they know that newborns are utterly helpless—fetal, in fact—and need their parents' love and care to create a "womb" outside the womb.)

Humans are simply born too early in their development, at the time when their heads will still fit through their mothers' birth canals. As babies' brains grow, during this extrauterine year of fetal life, so do their bodies. About the time of the infant's first birthday, the period of fetal brain growth terminates, coinciding with the beginnings of speech and the mastery of erect posture and bipedal walking. All of these defining characteristics of humans are developed or learned during the crucial first year of life. At this point, the human infant has reached a behavioral and developmental stage comparable to that reached by every other precocial mammal at birth. Discovering this has made me think differently about our first year. I am convinced it is the most important twelve months in our lives. The newborn's brain is growing, absorbing, sucking in impressions and information at a fetal rate; these experiences, in turn, affect the growth and development of the brain itself. Simultaneously, the baby's body is developing adult proportions and acquiring adult capabilities, such as the essential and complex abilities to walk and talk. Everything that happens—emotionally, physically, metabolically—during that first year must have a deep and lasting impact.

If humans deviate so strongly from the general mammalian pattern, what unique course do we follow in terms of brain growth? At birth, human brains are average relative to body size for a primate. Data on human conceptuses, embryos, and fetuses fall neatly on the line that shows the relationship of brain weight to body weight in all primates. But, unlike that of every other primate, our graph shows no point of inflection at birth; the line proceeds to rise, the rate of brain growth continues unchecked, until a full year after birth, when the slowdown occurs. The effect of prolonging this period of the fetal rate of brain growth is staggering: whereas every other primate roughly doubles its brain weight from birth to adult size, humans manage to more than triple theirs. I realized that this was another recurrence of the unusual human growth pattern: first we stay fetuses longer than we "ought" to, in order to grow our brains; next we stay children longer than we "ought" to, in order to keep learning and storing information in that big brain; then we are forced to undergo a dramatic growth spurt in order to catch up to the adult size we "ought" to attain. One demand of this strategy of growing brains after birth is that maternal nutrition must be reliable and good; both large bodies and large brains require regular intake of protein. Another requirement is a strong social organization, for how can a mother care for an unusually helpless infant (for a primate) and ensure that her diet is good without the help of others? Dubois was right; understanding the brain opens the door to many insights.

Because we had a complete braincase and a nearly complete pelvis for the boy, we could ask and answer an intriguing question: Had *Homo erectus* already evolved this uniquely human trait of prolonging the fetal growth rate after birth? Bob Martin had made a bold prediction in his publications on the subject. He speculated that hominid brains could get to an adult size of about 850 cc without developing secondary altriciality; bigger brains than that, he believed, would require the human evolutionary pattern to have evolved. I realized we could determine if he was right because our boy had a brain only slightly larger than his Rubicon: 880 cc.

The prediction was this: if the baby's brain that would fit through the boy's birth canal (suitably altered by a quick "sex-change operation" to make childbirth possible at all) was about

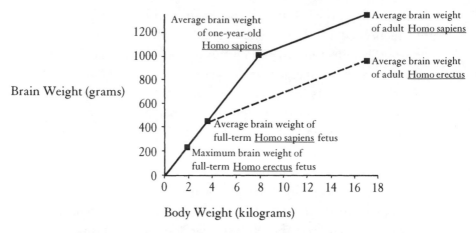

Brain Weight (grams)

Body Weight (kilograms)

After conception in humans, the brain grows very rapidly *in utero* relative to body size and reaches more than 400 grams at birth. Human babies' brains continue to grow at the fast fetal rate for the first year of life (solid line), more than tripling in size by the time adult weight is reached. In contrast, nonhuman primates slow their rate of brain growth after birth and do little more than double their brain weight between birth and adulthood. If *erectus* grew like a nonhuman primate (dotted line), then its newborn brain weight would be as big as or slightly bigger than that of a human newborn; such a brain would have been much too large to pass through a female *erectus*'s birth canal. Thus *Homo erectus* must have mastered the human evolutionary trick of giving birth to babies whose brains continued to grow rapidly during the first year of life.

half of the adult volume, or 440 cc, then *Homo erectus* behaved like any other primate and Bob was wrong. If the brain that would fit through the canal was close to but smaller than about 440 cc, then *Homo erectus* babies must have been among the first to be born "early" in developmental terms. This result would mean that Bob was right. If the baby's brain was much smaller than 440 cc, then secondary altriciality must have developed long before the 850 cc Rubicon and long before 1.5 million years ago.

The first task on the way to establishing *Homo erectus*'s reproductive strategy was to estimate the diameter of the bony birth canal. Chris Ruff and I undertook the project, which involved restoring the correct anatomical relationships among the elements of 15K's pelvis. Because of the boy's immaturity, the parts of his sacrum and two innominates (hipbones) were in several pieces. In life his pelvis, like his long bones, consisted of various bony pieces held together by cartilage growth plates, which would have been replaced by bone if he had reached physical maturity. But cartilage does not fossilize, so all that we recovered were the bony parts.

We used epoxy casts of the bones (to avoid any risk of damaging the original fossils) and a mixture of hot sculptor's paraffin wax and plaster of Paris to stick the casts together. Where the bones or fossils had been eroded or broken, we reconstructed missing segments out of the mixture by making a mirror image of the other side. Since an adult innominate of early *Homo* was known from Koobi Fora, and there was an adult female innominate of *Homo erectus* from Olduvai, we also consulted these specimens when we had to fill in missing areas. The boy's bones were so good that we couldn't be far wrong in our reconstruction; it was more science than art.

Once we had a reconstructed pelvis, we had to "grow" it to adult size. After all, the boy was only eleven years old at death, in human terms; he died too young to become anybody's ancestor in the literal sense. We also had to correct for the fact that he was male, not female. At this point, we stopped to ponder possible flaws in our procedures: we could err in "growing" the boy's pelvis or we could have had the wrong sex from the outset. We believed that the pelvis would enlarge as the boy grew up in the same way that human pelvises do, which seemed a reasonable assumption since his body proportions were in other respects so very human. The demands of thermoregulation, which we knew strongly affected the boy's body shape (and thus his pelvic dimensions), would stay constant as he grew older. We were also reasonably confident in our assignment of sex. Every measurement and analysis we could make suggested he was male and, reassuringly, analyses of some other *Homo erectus* innominates identified some who looked female. What matters during childbirth is not the external pelvic width but the internal, transverse (side-to-side) diameter of the bony pelvic inlet, where the baby's head engages. Because that diameter is about 4 percent larger in human females than in males, we were planning to increase the boy's measurement twice: once for growing to adulthood, and once for the sex-change operation that would turn him into a female. If we were wrong, and he was female to begin with, we would run the risk of enlarging his pelvis too much. It seemed an unlikely eventuality. We decided simply to acknowledge the possibility of error and reconsider its consequences after our estimates were completed.

The transverse diameter in the boy's reconstructed pelvis was about four inches, or 100 millimeters. Growing to adulthood like a human would increase this measurement to just more than four and a half inches (115.6 mm). To transform the now-adult pelvis into a female one, we had to enlarge it by another 4 percent, to four and three-quarters inches (120 mm). Before our boy would be ready to become a mother, we had to make one last adjustment: an allowance for the soft tissues that would have covered the bony pelvis in life. Based on data from modern humans, we calculated that the soft tissues would diminish the effective size of the pelvic inlet to four and one-third inches (110 mm).

The second task was to find out how big a brain would pass through this canal. A baby rotates as he or she passes through the birth canal. The head engages, or enters the pelvic inlet, facing to the mother's side so that the long dimension of the head is aligned with the greatest width of the pelvic inlet. Then the baby turns to exit the pelvis and usually is born facing its mother's back. So the crucial dimension is the front-to-back diameter of the baby's head, which must be smaller than the width of the pelvic inlet (or less than four and one-third inches, or 110 mm).

In human terms, this head size would belong to a fetus who is thirty-two to thirty-three weeks old, with a cranial capacity of about 192 cc. The maximum cranial capacity, by our reckoning, was 231 cc. This was a little more than half the estimate of 440 cc that would have indicated an apelike pattern of brain growth.

Now that we knew the size of a baby *Homo erectus* brain, we could estimate its subsequent growth trajectory fairly accurately, if it grew its brain like a human. A human baby's brain starts at 369 cc and increases, during the first year of life, to about 961 cc, after which the growth rate declines; adult cranial capacity is approximately 1,345 cc. So if *Homo erectus* started with 231 cc of brain, then on its first birthday it would have had just over 600 cc of brain, and as an adult its brain would have reached about 843 cc. This estimate is gratifyingly close to what we calculated originally as the boy's adult brain size; 909 cc represented only an 8 percent increase over 843 cc. What made the convergence of these estimates compelling was the fact that the calculation of 843 cc was based on a pelvic reconstruction while the 909 cc estimate of adult brain size was based on completely independent evidence from the skull, which already contained 880 cc when the boy died at eleven years old. The results confirmed that *Homo erectus* was secondarily altricial and the extreme narrowness of the *Homo erectus* pelvis meant that this human trait was established well before *erectus* evolved. Bob was right that an 850 cc brain required a period of postnatal growth at the fetal rate; he was wrong, however, in thinking that this was the Rubicon.

To try to establish just how large a brain could evolve without secondary altriciality, we calculated the effects of growing a 231 cc brain following chimp or gorilla trajectories too. The results were clear. If *Homo erectus* grew like a chimpanzee from an initial cranial capacity of 231 cc, then the adult brain weight would have been about 718 cc; if he grew more like a gorilla, the adult brain size would be even smaller, only 534 cc. These estimates of adult cranial capacity differed markedly from the boy's actual brain size (880 cc) *as an eleven-year-old.* At the time of death, his cranial capacity was already 22 percent larger than the adult brain size predicted by the chimp model and 66 percent larger than that predicted by the gorilla model. In short, it was impossible for him to have started as a baby with a 231 cc brain, followed an apelike pattern of brain growth, and ended up with 880 cc inside his skull when he died. These results suggested that all hominid brains larger than perhaps 750 cc were grown with the help of secondary altriciality, pushing the evolution of that uniquely human trait back into *Homo habilis* times.

The human evolutionary "trick" of growing a brain at fetal rates after birth was obviously well in place by 1.6 million years ago. In fact, it seems likely that the change in the pattern of brain growth from apelike to humanlike must have accompanied the evolution of *Homo habilis.* This highlighted the importance of body shape for thermoregulation or for bipedality, or maybe even for both; these demands were so great that the pelvis stayed narrow in early *Homo,* severely limiting the amount of intrauterine brain growth that could occur and forcing the development of secondary altriciality. Apparently the pelvis didn't enlarge to modern dimensions until later. When did that happen? We don't know what the pelvis of late *Homo erectus* looked like, but we do know that Neandertals, the archaic humans that succeeded *Homo erectus* in Europe and western Asia, had exceptionally broad pelvises. They also had Arctic conditions to contend with, so the need to dissipate heat through a long, slender body was obviated; in fact, their urgent need was to conserve heat, hence their stocky body shape and broad pelvis.

Our findings about brains and pelvises implied an additional characteristic even more evocative. Somehow, *Homo erectus* had managed to rear babies whose senses were going full tilt while their motor abilities were undeveloped. More than ever, this convinced me that *Homo erectus* was an intensely social creature, with strong cooperative ties to others of its species. I also knew that *Homo erectus* had faced and solved a major dietary problem, for nursing babies with rapidly growing brains demanded a significant upgrade in diet. Becoming an effective hunter is the most obvious way to increase the amount of protein in the diet—and protein is what is needed for big brains and big bodies. But before concluding how that protein was obtained, I needed to resurrect my analysis of the costs of becoming a predator, with this new perspective in mind.

11

THE PREDATORY HABIT

Acquiring the predatory habit had clearly changed the trajectory of human evolution. When had it occurred? In 1989 I decided to review the predictions I had developed several years earlier by comparing predators and herbivores and by looking at the adaptations of those two oddball species, the grasshopper mouse (herbivore-turned-carnivore) and the giant panda (carnivore-turned-herbivore). I had predicted that a newly evolved predator would be expected to show five attributes: (1) increased sociality; (2) lower population densities; (3) improved means of slicing up meat; (4) smaller gastrointestinal tracts with more given over to the small intestine; and (5) more free time.

I knew that sociality was a characteristic of *Homo erectus*. The care given to the diseased female, 1808, as she was dying of hypervitaminosis A showed that there were strong social bonds among individuals; so, too, did the fact that *Homo erectus* had secondarily altricial babies who would demand a lot of care. It was unlikely that *Homo erectus* babies could be parked in a nest or den and left by themselves: they would probably cry or yell, attracting hyenas or saber-toothed cats and other predators. Nor could a *Homo erectus* baby cling to its mother and come along as she foraged for food, because the babies were too undeveloped in their motor skills and the mothers probably did not have fur to hold on to. No, there must have been some sort of cooperative arrangement that encompassed both caring for young and obtaining food, and I knew from the large brains of *erectus* babies that the food supply was consistent and of high quality.

As for the population density of *Homo erectus*, there was good reason to suppose it was lower than in previous hominid species. Prior to *erectus*, all hominids had been confined to the continent of Africa, whereas *erectus* was found from Africa in the south to Java and China in the east, and less certainly in Europe. This pattern of distribution showed that there had been a radical expansion in the geographic range of *erectus* relative to earlier species; an inevitable consequence of geographic expansion would be a drop in population density.

The expansion itself was not particularly surprising, I suppose because humans are spread all over the world and paleoanthropologists have often succumbed to the temptation to cast human evolution as a story of increasing humanness. What had always puzzled me (and many other paleoanthropologists) was why this geographic expansion seemingly took so long to occur. Until recently, the oldest dates documenting the spread of *erectus* were from Java, where the oldest sites seemed to be about 1 or 1.2 million years ago, a conclusion reinforced by a less precise but comparably old date from the Israeli site of 'Ubeidiya. Dates from China were less than 1 million years, as were various poorly dated sites across Europe and into England too. In short, there was a good deal of rather sketchy evidence that *Homo erec-*

tus had been confined to Africa, like its predecessors, until about I million years ago. But if this geographic expansion was a consequence of (or a prerequisite to) the greater success of *erectus*'s adaptations, why didn't it happen nearly two million years ago, when *erectus* first arose?

In 1991 I attended a conference on *Homo erectus* at the Senckenberg Museum in Frankfurt that changed the facts I had been working with. Participants had been invited to view the *Homo erectus* fossils Ralph von Koenigswald had collected in Java between the wars, specimens now residing in the Senckenberg's collections, and to bring any casts or originals that they might want to compare to von K's material. I brought a cast of every part of the Nariokotome boy, which was a major attraction for my many colleagues who had not yet been able to see the original. At the "show and tell" session with the fossils and casts, I noticed an unassuming white-haired man whom I had never seen before. He spoke to few other participants but pored over the casts and original specimens with intense interest. In his right hand he carried a bronze-colored tobacco tin; within it, nestled in a bed of cotton, was a fossil mandible. From what I could see across the table, the specimen closely resembled 15K's jaw, so I picked up my cast of the jaw and approached him. I smiled and introduced myself; he replied that he was Leonid Gabunia, a paleontologist from Tbilisi in formerly Soviet Georgia. Broken French was our most fluent medium of communication, as I have no Russian and he has only a little more English.

When I showed him the boy's mandible, I could read the surprise in his face. In a moment, when he removed his specimen from the tobacco tin, I saw why: the jaws might have been twins, they were so alike. His mandible came from an excavation in the Georgian Republic town of Dmanisi, about eighty-five kilometers (fifty-two miles) southwest of Tbilisi and many thousands of kilometers from Nariokotome. To me, the most exciting paper at the conference was one Gabunia shared with an articulate young German woman, Antje Justus, who had actually excavated the jaw. He described the fossil, and she presented its geological context, dating, and associated faunal and archaeological remains. The fossil had lain in sediments above a fresh-looking lava flow that was dated, according to the team's first radiometric dates, to 1.81 ± 0.1 million years. Although they planned to refine this date by using more sophisticated techniques than those first applied, this preliminary work set a maximum limit on the fossil's antiquity. (Their later analyses yielded a date of 1.9 ± 0.2 million years for the lava below the jaw.) The jaw was closely associated with pebble tools, simple cobbles with a few flakes removed to create a sharp edge, like those in the earlier sites at Olduvai that were almost 2 million years old. Another clue was that the jaw lay directly underneath two exquisite saber-toothed tiger skeletons; they and the rest of the fauna suggested an age of at least 1.2 million years: a minimum limit on the fossil's antiquity. The chances were good that the Dmanisi jaw itself was between 1.6 and 1.8 million years old, roughly contemporaneous with the Nariokotome boy. While confirmation of this date would not make the Dmanisi mandible the oldest known *erectus*, it did make it the earliest known record of the species outside of Africa.

I gave Gabunia my cast of the boy's mandible on the spot, to take back to Tbilisi with him. I knew he would have difficulty obtaining a copy from overseas while his country was undergoing the political upheavals that followed the breakup of the Soviet Union. He was delighted and promised to stay in touch. Communications between us have been intermittent, hindered by political difficulties and economic travails, but Gabunia eventually managed to reciprocate by sending me a cast of his mandible. This fossil resolved my bewilderment about the long delay in the geographic expansion of *Homo erectus*'s territory by showing that the delay was an illusion produced by sparse evidence. The Dmanisi mandible shows that *Homo erectus* reached Georgia within a few hundred thousand years or less of the origin of the species (depending on the final dates for Dmanisi), a much more reasonable answer. Gabunia's publication of his results in an international scientific journal was delayed by the difficulties in his country, so for a few years the date of the Dmanisi mandible remained a sort of tantalizing rumor among paleoanthropologists. Having heard Gabunia and Antje Justus speak, and having examined the jaw for myself, I was convinced of the validity of the identification and the date, but many others adopted a wait-and-see attitude.

Then in 1994 a team in California now known as the Berkeley Geochronology Group published some unexpected new dates on Javan *Homo erectus* sites that bolstered the Dmanisi claims. These dates suggested that *Homo erectus* reached Java sometime between 1.6 and 1.8 million years ago. The research team that produced the new dates was led by Carl Swisher III and Garniss Curtis, two of the most technically able geochronologists in the world. Despite their competency, and despite the congruency between their results and the information from Dmanisi, these dates proved controversial.

Dating the Javan sites is difficult for a number of technical and historical reasons. The most secure dates come from regions where you can establish a sound stratigraphic framework for your fossils: a sequence of geological beds, interspersed with volcanic ones that bracket various fossils. The sequence ensures that at least the relative dates are correct; a date that appears to be out of sequence may call attention to technical problems. Since only the volcanic ashes, lavas, and volcanic glasses like obsidian—and not the fossils—can be dated, it is important to have dates both above and below each fossil (bracketing them) if possible. Even establishing a stratigraphic framework is incredibly difficult in Java, one of the most heavily populated regions of the world. There are almost no geologic exposures—areas where natural erosion has exposed the layers under the surface—because almost every square inch of land is covered in rice paddies.

The first serious attempt to date the Javan sites was undertaken by a Japanese-Indonesian team who spent from 1976 to 1986 walking through rice paddies and documenting every tiny exposure and even lowering themselves down wells on ropes in an effort to determine the stratigraphic sequence of the Javan sites. The team established that the oldest *Homo erectus* sites in Java were Mojokerto and Sangiran, the former having yielded a beautiful braincase of a child and the latter having yielded more crushed and fragmentary remains. The monograph and papers resulting from this project suggested that Mojokerto and Sangiran were no more than 1.2 million years old and might be as young as 730,000 years.

Geochronologists working in ancient layers like those that yield *Homo erectus* have several techniques at their disposal. The Japanese team had used fission-track dating, a technique that relies on the spontaneous decay of radioactive uranium-238 (^{238}U). The ^{238}U in volcanic crystals like zircon gives off particles that, when they decay, make tracks through the crystals. The number of tracks indicates how long the uranium has been damaging the crystals, which were formed during the volcanic eruption. Because crystals must be polished and etched before the tracks can be counted under a microscope and some conditions erase tracks, fission-track dates may underestimate the antiquity of the rocks. Under most conditions, fission-track dating is considered less precise and generally less reliable than the radiometric technique known as potassium-argon dating. Over time, potassium-40 (^{40}K) in volcanic rocks decays to stable argon-40 (^{40}Ar); the ratio of potassium to argon tells you how long ago the potassium was trapped in the rock. But because potassium is rare in Java, the ratio is very difficult to measure accurately and the K/Ar technique is unreliable for Javan sites.

Unsatisfied with the Japanese fission-track dates, Carl Swisher and Garniss Curtis turned to an elegant variation of K/Ar dating in which single crystals of volcanic mineral are irradiated, changing ^{39}K into ^{39}Ar, which is then released as the crystal is melted with a laser. The ratio of ^{39}Ar to ^{40}Ar, as measured in a mass spectrometer, then reveals the date of the volcanic eruption that produced the crystals. (The mass spectrometer separates the different argon ions by their differing charge and mass, and then records the quantity of each.) The single-crystal dating technique can be used on small samples; Swisher and Curtis applied it to obtain a date of 1.81 ± 0.04 million years ago for Mojokerto and 1.66 ± 0.04 million years ago for Sangiran, dates that were at least half a million years older than most people had believed. There was a sizable controversy, too, over exactly what the team had dated.

The crucial link in geochronology is the stratigraphy that connects the hominid fossil of interest to the volcanic eruption that was the source of the rock that has been dated. In this case, that link was a little weak, for the precise location of one of the two sites they dated, Mojokerto, has been unclear since von Koenigswald reported the fossil's discovery in 1936. The braincase was first announced as a surface find of the partial skull of a young child of *Homo erectus*, then still known as *Pithecanthropus* after Dubois's original finds. When Dubois—by then old, cranky, and argumentative—heard about the new skull, he cast aspersions on both its antiquity and its identity as a *Homo erectus*. Dubois asserted gruffly that the braincase was "really human" and had "nothing in common with *Pithecanthropus erectus*," which was not at all human. Von K let the matter drop (Dubois died a few years later), but his later publications told a different story of the skull's discovery. Von K described the skull as having been excavated from one meter below the surface by a Javan collector known as Andojo or Tjokrohandojo, who would prove to play a crucial role in subsequent events because his memory was the only record of where the find actually occurred. Various other publications by von K and other paleontologists described the Mojokerto site as being located at varying distances from several local villages, adding further confusion.

In an attempt to clarify the matter, Teuku Jacob, an Indonesian paleontologist, returned to the area with Tjokrohandojo in 1975, some forty years after the find had been made. The aged collector indicated a place and a layer from which the skull had been removed, which would have settled the matter were it not for the fact that in 1985 he told some Japanese researchers that a different layer had yielded the skull. After Tjokrohandojo died, and seventeen years after he and Jacob had visited the area, Jacob took Swisher and his colleagues to the spot that Tjokrohandojo had pointed out, so that the geochronologists could collect samples for dating.

Clearly, the flaw in this chain of evidence about the location of the find is that it relies too heavily on human recollection. The exact location of individual finds is extremely easy to forget, especially if the landscape has changed at all in the intervening years. Many paleoanthropologists working with written records and even sketch maps have been unable to relocate a fossil locality after the passage of even a year or two. And many collectors would sooner tell an esteemed visitor that the find was made *there* rather than say rudely that they cannot remember the spot.

Swisher thought he had found the perfect solution when he noticed the original matrix still stuck to the Mojokerto braincase; he could date it directly and not worry about the field site. He asked his Indonesian hosts for permission to remove enough matrix to use as a dating sample but, nervous about altering and possibly damaging the fragile specimen, they allowed him to remove only a tiny sample. The sample was too small for dating, but it included enough hornblende (a volcanic mineral) for

Swisher to run a "fingerprint" analysis. He compared the chemical composition of this hornblende with a sample gathered from the place where his team believed the Mojokerto cranium had been found. The results were suggestive but ultimately (and unfortunately) inconclusive. The hornblendes in the two samples were similar but not identical in the percentages of the various elements or chemical compounds. Oddly enough, the sample removed from the fossil was consistently a little more variable in composition than samples from the site. The dilemma is how to interpret these results. They certainly do not refute the hypothesis that the Mojokerto site has been correctly identified, nor do they confirm it. Unfortunately, no one knows how variable in composition hornblendes from a single eruption ought to be. But the resemblance between the samples left Swisher and his team confident that they had dated the right level at the right site. When their analyses yielded a similarly old date from a spot at Sangiran, a site whose location is not disputed, Swisher and his colleagues felt they had a strong confirmation of the correctness of their work.

Skeptics had a perfectly sound criticism, though: there was a missing (or at best damaged) link between the material that yielded the 1.81-million-year-old date and the hominid fossil. And there was the half-million-year discrepancy between the dates produced by Swisher's team and those from the earlier Japanese-Indonesian study. The offsetting advantage was that the Japanese-Indonesian researchers had dated a series of sites in a stratigraphic sequence—and the dates fell in the predicted order—whereas Swisher and his colleagues had two isolated dates, one of which might have been derived from the wrong site. While these criticisms are accurate, they do not address the heart of the matter: something in Java is 1.6 to 1.8 million years old, and it would make perfect sense if that something is *Homo erectus.*

There are some interesting implications inherent in these new Javan dates. On the face of it, they imply that the Mojokerto braincase is as old as, or maybe just older than, the oldest *Homo erectus* skull in Africa, which happens to be KNM-ER 3733, the *Homo erectus* skull from Koobi Fora that demolished the single species hypothesis. This implication led some adventuresome anthropologists to speculate that *erectus* might have evolved in Asia, not Africa. It is a fascinating idea, but one undermined by the lack of any plausible ancestors to *erectus* in Java or anywhere in Southeast Asia. There is not a scrap of a hominid until *erectus* arrives. In contrast, in Africa for two million years before *erectus* appears, there are thousands of specimens of several species of undoubted hominid that might be ancestral to *erectus*. The immediate predecessor to *erectus,* the troublesome *Homo habilis*, makes a fairly likely ancestor. It would take a huge number of new fossil discoveries in Southeast Asia to overturn this pattern.

In fact, although I find the Mojokerto and Sangiran dates plausible, I am not convinced that this makes the Javan fossils older than the African ones. First of all, the dates are on the rocks, not the fossils. When my colleagues and I say that 3733 is 1.78 (or about 1.8) million years old, we mean that it was found in a geologic bed that lies above a volcanic tuff dated to 1.89 ± 0.02 million years and below another dated to 1.64 ± 0.05 million years; interpolating either up from the lower tuff or down from the upper one suggests that the fossil is about 1.78 million years old. The Javan date of 1.81 million years has not been adjusted for the fossil's exact position because no one knows (or can know) where the fossil lay. And there is the matter of precision to be taken into account. These dates have margins of error that may incorporate as much as 400,000 years. Certainly any span of time shorter than a few hundred thousand years is invisible to our presently available dating techniques. Thus 1.78 million years in the African sequence may be older than, younger than, or even the same age as 1.81 million years in Java. Finally, 3733 is probably not the oldest specimen of *erectus* in Africa; it is simply the oldest skull of such completeness over whose attribution no one can sensibly argue. But there are some older postcranial bones that may be *erectus*, including a pelvis from 1.95 million years ago. The identification of this specimen is less certain than that of 3733. The pelvis matches the one from the boy, but because there is no known pelvis of the preceding species, *Homo habilis*, we cannot determine whether *habilis*'s pelvis was appreciably different from *erectus*'s.

Despite the controversies and arguments, the new dates from Java seem to reconfirm neatly the idea of a very early migration of *erectus* out of Africa, a once-faint notion that was revived by the Dmanisi mandible. In the last few years, the estimated age of 'Ubeidiya, the Israeli site full of Acheulian tools, has increased, giving that locality and the finds there a possible date of about 1.4 million years old. Perhaps, after all, *erectus* populations expanded rapidly soon after the species evolved. I think it is likely that they spread out of Africa via the Middle East to Eurasia, leaving their bones and tools at Dmanisi in the Caucasus and Mojokerto and Sangiran in Java at about 1.7 or 1.6 million years. It would be a mistake to think of this as a deliberate migration; the outward spread of *erectus* occurred in response to population pressures and the need to keep densities low rather than from a conscious decision to go north, or east.

The speed of the territorial expansion of *Homo erectus* is impressive, especially since geography, climate, fauna, and flora would be changing all along the way. It is the sort of expansion that predators undertake successfully—whether your prey is an antelope, a deer, or a goat is of little consequence so long as you catch it—while prey are more closely tied to particular plants and climates. In fact, prey animals rarely expand their territories so far or so fast without evolving into a new species. (If the timing seems implausibly rapid, remember that modern humans spread from Asia across the Bering Strait, through North America, and down to the southernmost tip of South America in no more than 15,000 and probably as few as 1,000 years.) This new

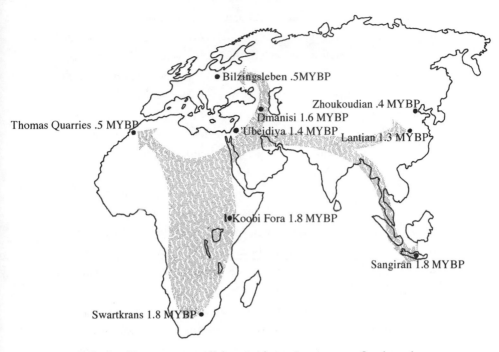

Prior to *Homo erectus*, all hominid species were confined to the African continent. But *erectus*, with its long legs, large brain, and strong body, spread rapidly throughout the entire Old World. This change in distribution is one of the signs that this species had changed its ecological niche, probably to become an efficient predator. (MYBP stands for million years before present.)

information supported perfectly my predictions of the consequences of becoming a predator.

Although sociality and reduced population densities were characteristic of *Homo erectus*, three of the predicted attributes were still unaccounted for. I expected to see evidence of improved technology for carcass-processing, more free time, and smaller guts. Primitive stone tools known as the Developed Oldowan industry were replaced at about 1.4 million years by the Acheulian industry; try as I might, I did not believe that either represented a major advance over the very earliest stone tools. Sharp rocks are sharp rocks, and any one of them will cut meat. The earliest stone tools preceded *Homo erectus* by 700,000 years or so; by a few hundred thousand years before *erectus*, stone tools were leaving cutmarks on animal bones, which means cutting up meat was not a novel problem for *Homo erectus* and using a stone tool on animal remains was not a novel solution. Hafting—attaching stone points to spears, arrows, or other sorts of "handles"—would have been a major advance, but everything in the Acheulian tool kit appears to be hand-held. It is difficult to envision how an animal can be killed using a stone tool that you have to hold in your hand; it is somewhat akin to the bizarre Victorian endeavor of killing a deer with a knife, which was only accomplished because Victorian hunters had the advantage of hunting dogs (which *erectus* surely did not). Maybe weapons that would not preserve well, like wooden spears, snares, or traps, were used to kill or immobilize animals; maybe it was the invention or the improvement of "invisible" tools like these that constituted the meat-processing advance I would have expected to see.

I was also unable to see evidence of increased free time in the fossil record, and initially I despaired of ever knowing anything about *Homo erectus*'s gut size. In the latter case, I was delighted to be proved wrong.

The first clues came from comparing reconstructions of the Nariokotome boy to those of the earlier australopithecines. Reconstructions of Lucy and her kin often have gorillalike potbellies. The fossils prove that Lucy's hips flared widely at the top (making for an exceptionally broad birth canal) and her reconstructed rib cage was shaped like a funnel, with the narrow part at the top and a wide lower region. This meant that her torso was thick-waisted, like that of an ape, and her gut probably protruded substantially. In contrast, 15K had a long, slender torso with a very human look to it. Lyman Jellema and Bruce Latimer of the Cleveland Natural History Museum joined me in analyzing the boy's rib cage in detail. We had at least one member of each pair of ribs starting with the first rib, at the top, and ending with the eleventh at the bottom. Only the twelfth ribs were entirely missing or so fragmentary as to be unrecognizable. After studying the anatomy, curvature, and declination of the boy's ribs, we concluded that his rib cage was indistinguishable from that of a modern human in almost every respect. It was entirely unlike the rib cage of a chimpanzee or gorilla (or Lucy). Like us, his thorax was barrel-shaped; like us, he must have had a well-defined waist between his narrow hips and his lowest set of ribs. In fact, because he had an extra lumbar vertebra by human standards, he might have seemed to have an especially long torso with a noticeably slim waist. This meant that he could not have the extensive large intestine that herbivores need in order to process their food; there was no room for it in his torso. We could prove that possession of a waist was not simply a function of the twisting of the torso during bipedal locomotion, for Lucy and the other australopithecines were already bipedal. It was just a question of gut size, and the boy's guts were small. Only predators can afford to have short gastrointestinal tracts, because animal foods are readily broken down, whereas the cellulose walls of vegetable foods have to be cracked open by cellulose-eating bacteria in the fermentation chambers of the gut before the herbivore can benefit from any nutrients.

Recently, two other colleagues put a different spin on the issue of gut size. Leslie Aiello of University College in London and Peter Wheeler of Liverpool John Moores University had intended to study the question of increasing brain size during hominid evolution from a new perspective. While I had asked what the costs were of becoming a predator, they had asked what the costs were of becoming brainy. Brains, as I have said before, are expensive organs; Leslie and Pete documented just *how* expensive, focusing on the metabolic budget. They uncovered some fascinating facts. By weight, the "running cost" of the brain—

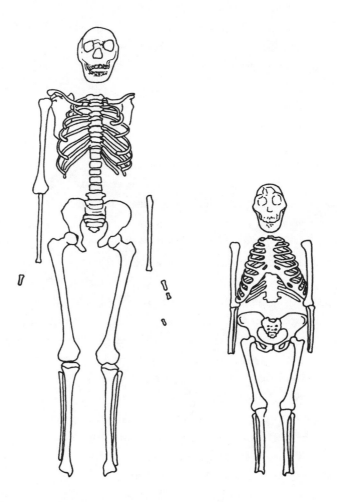

The partial skeleton of Lucy, a female *Australopithecus afarensis* (at right), shows that her figure was thick-waisted and potbellied, like an ape. In contrast, the Nariokotome boy (at left) was long in the torso and narrow in the hip, implying that he had a well-defined waist and a more human look to his body.

the basal metabolic rate, or BMR, of the organ itself—is nine times higher than that of a comparable amount of the whole body. Humans are, of course, exceptionally brainy animals; we have two and a third pounds (just over one kilogram) more brain than would be expected for an average mammal of our body weight. If humans have a lot of extra brain, and brain is metabolically expensive, then logically we ought to have a concomitantly higher total BMR—*but we don't.* Our total BMR is about average for a mammal of our size.

Leslie and Pete knew that some other tissue or organ system had to diminish in size to compensate for the increase in brain size, or else the metabolic budget for humans would never balance. (Unlike governments, bodies that run at a persistent deficit quickly cease to exist.) As they analyzed the size and metabolic demands of various tissues in the human body, they quickly perceived the answer: the human gastrointestinal tract is unusually small for a primate of our body size, by almost exactly the amount that is needed to compensate for the increase in brain size. This finding led them to articulate the expensive organ hypothesis (which might also be called the "brain versus brawn" theory): in order to evolve larger brains, hominids had to evolve smaller guts.

When they hear this explanation, some people leap to the incorrect conclusion that humans with bigger guts are therefore less intelligent, or even less brainy. As Leslie and Pete point out, the relationship is a broad-scale one that applies across species, not to the relatively minor variations within the species. Besides, as a person thins or fattens, his gut does not lengthen or shorten; it is only that his girth—"gut" only in the colloquial usage— changes. However, Leslie's and Pete's conclusions imply that braininess probably coevolved with predatory behavior and that a relatively encephalized species, like *Homo erectus*, was almost certainly an effective hunter. As Bob Martin observed, a high-quality diet is essential for a mother nursing a big-brained baby; Leslie and Pete would add that such a diet is also imperative for a big-brained mother or father simply to maintain herself or himself on a daily basis. Brains and diet are inextricably intertwined, for the Nariokotome boy one and a half million years ago and for us today too.

12

A BALANCED PERSPECTIVE

Brains and diet were an important part of 15K's ecological adaptations, but so was his way of moving about the world. We had known, of course, that *Homo erectus* walked upright. Because Eugène Dubois's first momentous discovery in Java in 1891–92 included both a skullcap and a femur, this was one of the few facts that had formed an integral part of the image of the species, and had dictated its scientific name as well. While it is obvious that the major bones of the leg and foot must be adapted to bipedalism, the evolution of this new mode of locomotion also produced subtle changes in other parts of the body in order to maintain balance and equilibrium during bipedal walking or running. Put plainly, bipedalism is a precarious and difficult way to move around the world. Ask any infant. There is a good reason why many quadrupeds are able to get up and move like an adult within minutes of birth, while human babies take a good year or so to start toddling unsteadily about the world under their own steam. Any quadruped normally has at least two feet on the ground while the others move forward, whereas a biped must pivot over a single support while the other leg swings forward. From an engineering perspective, bipedalism is a ridiculous answer to the need for locomotion, posing problems akin to balancing an apple on top of a moving pencil.

One of the keys to habitually erect posture lies in the trunk, which plays a big role in the balancing act. It is the muscles of the torso, working on the skeletal framework of the vertebrae and ribs, that help keep us balanced over the supporting, or stance, leg during walking. The spinal column is especially crucial, so I asked my former student Carol Ward to study the vertebrae of the Nariokotome boy. She is an expert in this region of the body, having done a fine analysis of the anatomy of the lower back of a fossil ape for her Ph.D. She decided to collaborate with Bruce Latimer, who had analyzed the ribs with me, and together they set about looking for changes from the primitive apelike condition. How the vertebral column had been reshaped in response to the demands of two-footedness would give us a new perspective on the extent of the boy's adaptations to upright posture.

Because apes, like gorillas and chimps, are quadrupedal most of the time, their vertebral column has a different task to perform than humans'. Our spine is weight-bearing; the double S-shaped curve in our vertebral column—curving forward through the neck, back through the thorax, and forward again through the lumbar region—helps absorb the impact of walking by acting like a spring. Because humans habitually walk and stand upright, the curvature also helps by balancing our body weight, with shoulders, hips, knees, and ankles in vertical alignment. This curvature is produced in part by the wedged shape of some of a human's vertebral bodies, the columnar portion of each vertebra that, along with the intervertebral disks, actually bears weight. If the upper and lower surfaces of the vertebral body were parallel, then the spinal column would be straight and vertical. However, in the thoracic region, the vertebral bodies are thinner at the front and thicker at the back, making a curvature that is concave when viewed from the front. In the lumbar region, the reverse is true; vertebral bodies are thicker toward the front of the body and thinner at the back, making a convex curvature when viewed from the front. The extent of wedging is measured as the angle between the upper and lower surfaces of the vertebral body. Because apes' spines do not carry the full weight of their bodies during normal locomotion, their vertebrae show a small amount of wedging only in the thoracic region; this produces a very mild curvature that is concave when viewed from the front.

Apes also normally differ from humans in the number of vertebrae. Roughly 96 percent of the human population has twelve thoracic vertebrae (one for each pair of ribs) and five lumbar vertebrae. The great apes (gorillas, chimps, pygmy chimps, and orangutans) of Africa and Asia have thirteen thoracic vertebrae and three or four lumbars; the lesser apes of Asia, the gibbons and siamangs, have thirteen thoracic vertebrae and five lumbars. In Carol's previous work on the spine of *Proconsul*, an ape from Kenya that is about twenty million years old, she had found that it probably had six lumbar vertebrae. Since *Proconsul* can be taken as a reasonable model for the ancestor of modern apes and humans, Carol proposed that the primitive condition for those groups was to have a high number of lumbar vertebrae. As might be expected, Old World monkeys like baboons and vervets also have many lumbar vertebrae, normally seven. Presumably, through the course of their evolution, different modern groups reduced the number of lumbars by different amounts. The greatest reduction evolved in great apes. This situation would increase the stability of their lower backs at the expense of flexibility. Lesser apes have retained a more primitive, and more flexible, configuration, as have modern humans, but both would have less flexibility and more stability than their ancestors. Humans, with their uniquely upright posture, have also evolved vertebral bodies with a large surface area, another adaptation for weight-bearing and stability.

Carol and Bruce wanted to discover the extent to which 15K showed the structural and mechanical adaptations to bipedal walking seen in modern humans. Although the vertebral column is not complete above the pelvis—there are parts of only nineteen vertebrae preserved, whereas humans usually have twenty-four—it is the most complete column known for an early hominid.

The wedged shapes of the vertebral bodies showed clearly that the boy's spinal column had a double S-shaped curve, just like our own. All the anatomical details of this adaptation to weight-bearing were also present. Yet Carol and Bruce noticed two fascinating differences from modern human anatomy. First of all, they confirmed my observation that parts of six, not five, lumbar vertebrae were preserved in the boy's skeleton. This is a more primitive arrangement, more like that seen in *Proconsul,* than they had expected, but a small number (fewer than 4 percent) of perfectly normal humans have six lumbars too. But when Carol and Bruce went to South Africa to study the few vertebrae from a specimen of the even more ancient hominid species, *Australopithecus africanus*, they confirmed for themselves that this specimen, too, had six lumbars, as John Robinson had reported. The chance that the only two specimens of early hominids that preserve the lower spinal column would both sample a very rare condition seems very slight.

Still in South Africa, my two colleagues also saw a third specimen in which the lumbar vertebrae are preserved, but the specimen was not then published (nor is it yet). This is a touchy situation. Professional ethics dictate that the right of first description or analysis belongs to the team who found the fossils, or another scholar whom they designate. After all, the excavators have gone to the trouble of organizing an expedition; obtaining funding and permits; and documenting, cleaning, and curating the specimen properly. To expect first access to the fossil as compensation for this enormous effort is only reasonable. But sometimes publications take years to appear, depending on the difficulty of the analysis and the other commitments of the scientists involved. As a courtesy, sometimes a visiting colleague (especially one who has traveled a long way) is permitted to examine unpublished material on the understanding that no public comment or publication will mention the material until it has been properly published. While waiting years to discuss a specimen that has altered your perspective crucially may be infuriating, such agreements are vital. Otherwise fossil finders are faced with the choice either of being secretive and refusing to allow colleagues access until a publication is out, or of running the risk of being scooped on their own material. Carol and Bruce, both of whom have excavated fossils themselves, understand this principle thoroughly, so they have been very circumspect about what they saw. But if this specimen, too, should have six rather than five lumbar vertebrae, it would push to the breaking point the probability that this is a rare condition. The conclusion would have to be that six lumbar vertebrae was the normal condition for early hominids, a dramatic discovery with interesting implications.

Carol and Bruce mapped out a scenario for the evolutionary reduction of the number of thoracic and lumbar vertebrae from the primitive condition of thirteen thoracics and five lumbars. As posture and locomotion changed in the hominid lineage, the lowest thoracic vertebra evolved into an additional lumbar vertebra; at the same time, the lumbar vertebrae began to develop the wedged shape that creates the lumbar curvature, which is called lordosis. Evolving a humanlike degree of lordosis improved the postural control of the torso in a habitually erect biped. Because early hominids had an extra vertebra in their lumbar series, the same lordosis could be produced with a smaller change in the shape of the body of each individual vertebra than would have been the case with only five vertebrae. Only the lordosis was present, so apparently the number of lumbar vertebrae was reduced further as selection favored additional stability.

The only other difference between the boy's vertebral column and that of modern humans is that the surface area of his vertebral bodies was unusually small for his weight. In this regard, he was more apelike, but apes do not have to support their body weight through their vertebral column in all postures and movements. Again, Carol and Bruce found this same pattern in the South African specimen of *Australopithecus africanus*, so they concluded that this was a primitive characteristic that had been retained. At some later point in our evolutionary history, perhaps at the same time that the number of our lumbar vertebrae was reduced from six to five, selection for greater stiffness or stability of the torso must have acted to favor a broader weight-bearing surface.

At this point, I thought we had a balanced perspective on the adaptations of *Homo erectus* to bipedalism. We knew he was long-legged and slim-hipped, both adaptations that would improve the efficiency of bipedalism, and showed substantial changes in the shape and number of his vertebrae to favor stability of the back in an upright position. It was only in 1993, when I read the draft of the Ph.D. thesis of a young Dutch anatomist, Fred Spoor, that I realized there was another perspective I had overlooked.

The insight that lay behind Fred's research is that another essential component in any locomotor pattern involves the organ of equilibrium, also known as the organ of balance. This organ literally provides your sense of balance regardless of the position you are in. It also controls an important ocular reflex that lets your eyes track as you move, so that the ground (or tree limb or any other substrate) is perceived as an unblurred, clear surface even while you move. This ocular reflex can be demonstrated by a simple experiment. If you jiggle the book you are reading rapidly, you can no longer hold the words in focus and read; the letters become an uninterpretable jumble. However, if you hold the book steady and wiggle your head rapidly, you can continue to read without much difficulty. This is because, as you move your head, the ocular reflex signals your brain, telling it exactly where your eyes should look so that it can interpret appropriately the information they send to the brain. The evolutionary point of this ocular reflex is not to make reading easier; it is to enable an organism to be quite sure where its feet are about to land even (or especially) when you are moving rapidly. Thus, any change in locomotor pattern (and particularly in the rapidity and predictability of movement) must be accompanied by a change in the organ of equilibrium.

The hardware for the organ of equilibrium is a complex of membrane, fluid, and nerves, known collectively as the vestibular system, that lies deep within the skull in the inner ear. The housing for this intricate structure—all that is preserved in fossils—is known as the bony labyrinth. Part of it is a snail-shaped structure known as the cochlea, which houses the organ of hearing. But the rest of the labyrinth is comprised of a chamber known as the vestibule and three tubes called the lateral, anterior, and posterior semicircular canals; these canals loop through the petrous bone of the skull. The semicircular canals are arranged at rough right angles to one another and are lined in life with a fluid-filled membrane. Special cells situated at the ends of the canals sense position and movement and are also the sensors for the ocular reflex. On theoretical grounds, it has been shown that the sensitivity of the vestibular system to posture or movement is tuned by the canals' dimensions, particularly by the radius of curvature of each canal. A somewhat obscure paper published early in the twentieth century confirmed this relationship between structure and function, showing that fast-flying birds had bigger canals than slow-flying or ground-dwelling birds. When Fred read this publication, he realized that the size of the semicircular canals might be used to deduce the habitual patterns of movement of extinct mammals, especially if the locomotor pattern were unusual. He decided to find out if the inner ear could reveal anything about the controversial matter of the timing and evolution of hominid bipedalism.

The bony labyrinth is not an easy subject to study. The entire structure of the inner ear is, in most mammals, less than an inch in size, and it lies within a particularly dense piece of bone known appropriately enough as the petrous, a name derived from the Greek *petra,* meaning "stone" or "rock." It would be unrealistic to hope that fossil skulls would be broken so that the semicircular canals are visible and undamaged, and even more unrealistic to think that any curator would allow his or her precious fossils to be sliced up like a loaf of bread. (This procedure, known as serial sectioning, is the classic method for studying the inner ear of modern species.) But, with the help of an imaging expert, Frans Zonneveld, Fred was able to use computerized tomography, or CT scans, to examine fossil and modern primate skulls without inflicting any damage. CT scanning is an advanced technique for taking a series of X rays of a subject (usually a patient), gathering a series of radiographic slices that can be reconstructed into a three-dimensional whole with the aid of computers. Although the radiation dosage has to be adjusted for scanning the skull from a monkey skeleton or a fossil hominid, rather than a living human, Fred was able to obtain exquisitely accurate information about the vestibular systems of his long-dead "patients."

As is often the case in innovative research, his first task was to build a frame of reference within which his results could be placed. Fred painstakingly scanned almost one hundred specimens of twenty different living primates. This reference set ranged from small bodied to large, from primitive prosimians like lemurs to monkeys, apes, and humans; it included species with a wide range of locomotor patterns. He was able to gather additional information from the literature to round out his sample. All of this preliminary work established that he could obtain accurate measurements from CT scans of this type and, as he had expected on theoretical grounds, that the different types of locomotion used by these various species are reflected in the dimensions of the semicircular canals. With such a varied reference sample, some differences in vestibular dimensions were simply an effect of variations in body size; these could be factored out mathematically. Once this correction had been applied, it was clear that fast-moving, agile species, like long-armed gibbons that swing through the trees, or big-eyed, fast-leaping tarsiers, have consistently larger semicircular canals than do slow-moving terrestrial or arboreal species. The data on humans stood out because we stand, walk, and run upright with a frequency matched by no other living primate. This locomotor peculiarity has shaped our vestibular system into a unique configuration. Relative to great apes like the chimpanzee or gorilla, humans have larger anterior and posterior canals and a smaller lateral canal, an arrangement well suited to monitoring movements that occur in a predominantly vertical plane.

Once Fred had perfected his methodology and analyzed his reference sample, he turned to the fossil record. Because he was pioneering a new approach to these questions, Fred didn't have enough time, money, and access to study every fossil hominid he might have wished for. He chose to examine a series of twenty hominid specimens mostly from South Africa, including *Aus-*

tralopithecus africanus, A. robustus, Homo habilis, and *H. erectus.* His sample included skulls that many anthropologists would arrange in a single lineage: *Australopithecus africanus* to *Homo habilis* to *Homo erectus,* with specimens of *A. robustus* representing an evolutionary side-branch roughly contemporaneous with *A. africanus.* (My phylogeny is shown in Figure 5, on page 150.) A few parts of our ancestry are fairly clear. I consider *Australopithecus afarensis* to be probably ancestral to all later species of australopithecines (but not necessarily to any species of *Homo*) and *Homo habilis* (whatever it may be) to be ancestral to *Homo erectus* and in turn to *Homo sapiens.* But the relationship between the two oldest and most recently discovered species, *Australopithecus anamensis* at about 4 million years and *Ardipithecus ramidus* at 4.4 million years, is very unclear at present, as is the relationship between these two early species and the later hominids. The obvious hypothesis, that *Ardipithecus* gave rise to *A. anamensis,* which in turn evolved into *A. afarensis,* and so on, is possible, and simplicity and chronology recommend it. Time will tell. However the species are arranged into lineages, the australopithecine specimens represent the vestibular apparatus in a genus *some* member of which gave rise to *Homo habilis, H. erectus,* and ultimately, ourselves.

The first question was when the modern vestibular adaptations for full bipedalism had appeared in the evolutionary lineage from apelike ancestors to humans. The first answer was: *Homo erectus.* This was an enormously satisfying answer to me, since every analysis we had conducted on 15K suggested he was fully bipedal in a modern sense. Even though Fred has yet to scan the Nariokotome boy's skull, I am sure the results will confirm his findings on other specimens.

Learning that *Homo erectus* had a modern vestibular system was not as interesting to me as the result of Fred's scan of a specimen known as StW 53, a small skull attributed to *Homo habilis* because of its resemblances to OH 13 (Cindy) and 24 (Twiggy).

It also seems very similar to OH 62, Don Johanson's partial skeleton of *Homo habilis*. Now I know Fred is a fairly conservative person, not one to relish the role of iconoclast unless he is quite sure of his information, so I listen carefully when he finds something unexpected or unusual. And the information his scans yielded about the inner ear of *Homo habilis* is certainly provocative in the context of an evolutionary sequence that starts with an australopithecine and ends with *Homo erectus*. I have already said that *Homo erectus* looked like a modern biped in terms of its inner ear anatomy. At the starting point of this sequence, the four individuals of *A. africanus* that he examined were rather similar to great apes in their size and morphology. So were the five specimens of *A. robustus*, although there were a few indications that both of these species had evolved in the human direction from an apelike condition. Because their bony labyrinths were more apelike than human, this finding lends support to those who think that bipedalism in early hominids (before *Homo erectus*) was not fully modern. Although it has been argued that the anatomy of these species indicates a tree-climbing component to their locomotion, nothing about the size or shape of the vestibular apparatus reflects arboreality or terrestriality per se.

Although Fred might be criticized for not having scanned the earliest hominid species, neither *Ardipithecus ramidus* nor *Australopithecus anamensis* had been found at the time of Fred's study. Their omission makes the situation clearer in any case. If *ramidus* or one of the early australopithecuses, *A. anamensis* or *A. afarensis,* has a rather apelike vestibular system, as you might expect, then the old argument about mosaic evolution can be raised. If one or all have an apelike bony labyrinth, then this may be nothing more than an evolutionary holdover, a trait retained until all parts of the body have time to "catch up" to the new locomotor pattern, bipedalism. At this writing, the earliest certain biped was *A. anamensis* at about four million years. Since *A.*

africanus and *A. robustus* lived more than one million years and maybe as many as two million years later than *anamensis,* sure enough time had passed for this new locomotor pattern to become established and for the organ of balance to evolve to accommodate it. No, the morphology of the inner ear of australopithecines is not an anatomical system "in transit," and the retention of an apelike morphology bespeaks the retention of an apelike component in the locomotor system.

Finding that australopithecines were still apelike in their inner ear morphology creates the inevitable expectation that *Homo habilis* would be intermediate between the australopithecines' apelike condition and *Homo erectus*'s modern one. This "in-between" position has always been assigned to *habilis,* no matter what part of the anatomy was being discussed. Fred's research strongly reinforces the message of the analysis of *habilis* postcrania by Sigrid Hartwig-Scherer and Bob Martin in showing how inadequate this idea is. Fred's scans showed that the inner ear of *H. habilis* looks more like that of a gibbon or even a monkey than either a great ape or a human. "It's very difficult to interpret," Fred says of the enigmatic morphology of *Homo habilis*'s inner ear. "The only thing that the labyrinth suggests is that it is less bipedally adapted than the australopithecines." But he can see only two possibilities: "Either this specimen is not *Homo habilis,*" despite its palatal resemblance to OH 13 and 24, "or, if it is, *Homo habilis* is unlikely to be ancestral to *Homo erectus.*" I couldn't agree more.

Fred's work has been controversial, partly because people are always reluctant to endorse findings based on a new approach until they understand it thoroughly and partly because, I think, the concept of *Homo habilis* as the single intermediate species between australopithecines and *Homo erectus* is so appealingly neat and tidy. I don't think there can be much question that inner-ear morphology reflects locomotion, so I wouldn't doubt his results on that basis, nor can I fault them because they indicate that a good, well-known hominid isn't very hominidlike. I have long felt that *Homo habilis* was a poorly defined species that encompasses specimens with too great a range of morphology to be grouped together, and Fred's work adds yet another discrepancy.

One of the exciting things that is bound to happen in the next few years is that Fred will be able to scan additional specimens of fossil hominids. I predict that when Fred scans the skulls of large habilines, like 1470, he will find them different from and more human than those of at least some of the small habilines. The rare large-bodied and large-brained creatures like 1470 and 1590 look to me like something that could have been ancestral to the Nariokotome boy and other specimens of *erectus*. The false pretenders to the throne, I think, are those small-brained specimens like 1813, OH 13 and 24, and StW 53. Whatever these represent, it is a strange ape-bodied, hominid-headed species that probably went extinct without issue, like so many other early hominids. When we find out what those small habilines were truly like, I think we will be forced to wrestle anew with the enormous identity question.

What does it means to be a hominid? What makes this one an ape, that one a human? It is not any huge genetic difference, that we know for sure. Identity must lie in the details, for we have the same number of arms, legs, fingers, toes, and teeth as apes; our bodies are built to a common plan. Yet neither I nor anyone else has much difficulty in telling apes from humans, alive or dead. It seems so simple. How big is your brain, how large your jaw, how complex your tools, how bipedal your adaptation? On one side of the divide stands the ape: furry, quadrupedal, smaller brained and bigger jawed, largely tool-less. On the other stands us and *Homo erectus*: hairless, upright, big brained and small faced, maker of lasting tools. But maybe there is no dichotomy at all, just one long, labyrinthine continuum full of evolutionary experiments and unthought-of combinations of humanlike and apelike traits. In a real sense, the "missing link" is an artificial construct and an unholy grail. I am not searching the Kenyan desert for some mythical chimera that lies between apes and humans. I am looking for the truth: about us, about them, and about our similarities and differences. I am striving to see the human animal in the right perspective.

13

RENDERED SPEECHLESS

These findings were fascinating, but I was feeling intellectually itchy, bothered by a nagging feeling that there was something big about the boy that we hadn't yet discovered. The idea hovered at the threshold of consciousness, an intuition that could neither be articulated nor dismissed. It was, I suppose, based on an observation I had made without even realizing it that nonetheless demanded explanation.

Even during those amazing days when we were excavating 15K's skeleton, I had noticed that his vertebrae were very odd in a way that lay outside of Carol and Bruce's purview in their study of adaptations to bipedalism. What I had seen was that the "hole" through each vertebra, known as the vertebral foramen, was rather narrow. This is not a trivial anatomical detail. Since vertebrae rest one on top of the other, separated in life only by the intervertebral disks, the stack of "holes" enclosed by the bony arches makes a bony tunnel, the vertebral canal, into which the spinal cord fits. The spinal cord is the home for the nerve fibers that control most of the body; many of the nerve cell bodies lie in the spinal cord too. If the canal is narrow, the spinal cord is small, meaning that the boy either received less information from his senses about the world around him or that he was less capable of a finely tuned response to that world. Anatomists designate these two types of information as either motor, impulses from the brain to the body that produce movements, or sensory, impulses that travel from the body to the brain to describe what has been encountered. A quick and dirty comparison showed that the canal in the highest of the boy's vertebrae that we possessed (one that came from the lower end of the neck) had only about one half the area of the canal at a corresponding point in a modern human skeleton. I was surprised momentarily, but then realized that I should have been able to predict this finding. After all, the brain and the spinal cord form an integrated unit, the central nervous system, and I already knew that his brain was only about two thirds as big as that of modern humans; it was entirely consistent that the spinal cord would be small too. All I could deduce from this observation was that he must have been an awful klutz. As an afterthought, I wondered if this was why the Acheulian tool kit was so repetitive and simple; maybe he and other *Homo erectus* individuals were just not coordinated enough to make anything better.

I knew, though, that I needed an expert to tell me what this diminished spinal cord implied, so I called on Ann MacLarnon. Ann is a calm, competent biologist at the Roehampton Institute of Whitelands College in London; she is intelligent and insightful, and I knew she hadn't the temperament to go making any wild claims without a firm basis for doing so. Ann's main credential for this task was that when she had been a graduate student she had done an extremely thorough comparative study of the spinal cords of a whole series of different primates. She had a unique data base of measurements of vertebrae and spinal cords, down to estimates of the number of cell bodies of motor and sensory nerve cells in the spinal cord at different points. Ann was just the woman I needed to look at the boy's vertebral column as a whole and to tell me something about what he could and couldn't do.

With the help of a colleague, the next time I was in Nairobi I took the measurements Ann needed from 15K's vertebrae and sent them off to her. Then I settled back to wait for her results. I received a draft of a manuscript some months later, which I read through eagerly.

Ann's previous studies had shown that the dimensions of the vertebral canal are an excellent indicator of the width of the spinal cord itself, except at the pelvic end of the body, where the spinal cord peters out. In most primates, the dimensions of the spinal cord reflect the body size of the animal itself: big animals need more nerve tissue to maintain control over their bodies. Since all of that nerve tissue ultimately comes from or goes to the spinal cord, it is no wonder that big-bodied primates have big spinal cords and small-bodied ones have smaller ones. Ann had also discovered that the main difference between humans and all other primates was an enlargement of the human spinal cord in the region that controls the lower neck, arms, and thorax, which, fortuitously, was the very region that was best represented in the boy's remains.

Ann confirmed that the boy's spinal cord was genuinely small in the thoracic region, as I had suspected. That made him anatomically like apes and monkeys and unlike humans. The lingering question was *why*, why humans developed this unique expansion of the spinal cord and what was implied about the boy's behavior by the fact that he didn't show a humanlike expansion. Typically, Ann took a very empirical approach to answering the question. She had dissected dozens of primates, including humans, and had studied the composition of the spinal cord in this region. Spinal cords contain two types of tissue, known as gray matter and white matter for their appearance. The parts of the spinal cord that house nerve cell bodies, where the nucleus resides, appear gray; the long nerve fibers, covered with fatty, myelinated sheaths, look white. Ann found that the enlargement of the spinal cord in the thoracic region of humans was due to an increase of gray matter, meaning that there were extra nerve cell bodies in that region. This made perfect sense; the extra cells showed that there were extra spinal nerves that left the spinal cord in this region. The surprise was that the location of the cell bodies told her that the spinal nerves were not for control of the muscles of the arms. (Although control of the arms is extremely important to us, it is just as crucial to nonhuman primates, who both manipulate objects habitually and use their forelimbs for locomotion.) These extra nerve cell bodies in humans reflected extra nerves to the abdominal and thoracic muscles. Ann's manuscript simply made this observation without drawing any further conclusions. I was dissatisfied and wrote back, pressing Ann to draw out the broader significance of her findings. What did it mean that humans had additional innervation of the thorax and abdomen? Why did we need it—and why didn't the boy need it?

She replied that she could think of only two interpretations of these data. The first is that *Homo erectus* was not yet fully adapted to bipedalism in terms of postural control of thoracic movements; by this she meant the twisting of the torso that inevitably accompanies walking, the characteristic movement that makes us swing the right arm forward with the left foot, or the left arm forward with the right foot, for balance. *I can't believe that, Ann,* I thought as I read her letter. She knew as well as I did that hominids had been walking for more than two million years by the time the boy was born. There was little chance that he and his conspecifics were still ill-adapted to bipedalism so long after this mode of locomotion had evolved. Besides, the work that Carol and Bruce had done showed that the vertebrae were almost fully adapted to weight-bearing. Since this is one of the most fundamental anatomical responses to bipedalism, I was incredulous that *erectus* could have adapted so fully to bipedalism in that respect—not to mention the reshaping of the pelvis, knee, foot, and (as Fred's work would later show) vestibular system—and yet still lack the nervous control of the muscles of the thorax. It made no sense to me.

Ann also offered an alternative explanation. "I gave a seminar in my department covering a lot of my spinal cord stuff, and the analyses I'd done on WT 15000 so far," she wrote. "A colleague, Gwen Hewitt, suggested that the increase in thoracic innervation in modern humans might be the result of increased breathing control associated with the evolution of speech." In other words, the extra nerve cells controlled the intercostal and abdominal muscles of the thorax. I quickly reviewed in my mind what I knew about these muscles. The intercostals are a set of muscles each of which runs between one rib and its neighboring (higher or lower) rib, in an arrangement that resembles the webbing between the toes of aquatic animals. These muscles help the rib cage work as a coordinated unit in breathing, so that all the ribs rise and fall, and move outward and inward, together. The intercostal muscles also contract every time you breathe in and out; they keep the wall of the chest firm so it doesn't balloon outward like an air mattress. Overall, then, the intercostal muscles function to coordinate and control inspiration (breathing in) and expiration (breathing out) very precisely. Abdominal muscles have a similar function of maintaining the integrity of the body wall during breathing. Because babies have such small chests, their abdominal muscles are used much more than the intercostals during breathing. Professionally trained singers usually breathe abdominally, too, as a means of increasing their lung capacity and increasing their fine control over expiration.

I considered the implications of Ann's (or Gwen's) idea. It is obvious that human speech is more than just making isolated sounds, which any animal can do. Humans have to get the intonation and the phrasing of the sentence right as well as the pronunciation of the words; otherwise that funny mechanical voice produced by computers emerges. If *Homo erectus* did not need the innervation to control breathing properly, that implied that the boy could not talk.

I told myself that this explanation could not be correct, either. We *knew* the boy could talk; he had a clear Broca's area. That was the bump on his brain that formed the hollow on the left side of the braincase, the one into which I had placed Samira Leakey's finger that long-ago day in camp, when her mother and I were first gluing the skull together. Ann's answers left me restless. I could sense that something important was eluding us. For the next half hour, I puttered around my office distractedly and then went into the lab to interrupt all the students and ask them what they had learned that day. Nobody had discovered anything interesting enough to take my mind off Ann's work, so I explained her ideas to them. No one had a good idea about the intercostals or enlargement of the spinal cord, either.

I wandered out again, and as I passed the bulletin board in the hall, I scanned it as a matter of habit, looking for seminar and lecture notices. *Maybe I could go listen to someone else talk about something interesting and forget about the Nariokotome boy for an hour,* I thought. I found what I was looking for, unconsciously: Marcus Raichle of Washington University at St. Louis was speaking in just a few minutes about positron emission tomography (PET) scanning. I am fascinated by new technologies, and I know new technology drives new ideas. Since I knew very little about PET scans, I went along happily.

PET scanning is a major new tool for learning about brain function. The patient is given a small dose of a very short-lived radioactive isotope into his or her bloodstream. While the isotope decays, the patient's head is scanned in a series of horizontal "slices," using a special sort of radiation detector. Raichle's group administered the isotope to normal medical-student volunteers and then gave them tasks to perform. As their brains worked, the blood flow was increased to the regions that were active; specific areas that commandeered a lot of blood showed up as bright white spots on the scan, while progressively less active areas were red, orange, yellow, green, and finally blue. Resting areas were black. The scans actually made the working of the brain visible as the task was performed over a matter of milliseconds. The work by Raichle and his colleagues was an elegant use of technology.

One of the first things Raichle's group decided to investigate was language. He exposed the students to a word, either by flashing it on a screen in front of their eyes or by pronouncing it over a sound system. The PET scans allowed him to trace the way in which words were taken in and processed and also to record the differences between hearing a word (auditory input) and reading a word (visual input). The next task for the students was to read (or hear) a word and repeat it, thus highlighting the areas of the brain responsible for producing speech. According to traditional ideas, Broca's area in the left temporal region functions as a word-generator and should light up just before the word is repeated; this should be followed by the regions of motor cortex on both sides of the brain that are responsible for firing the muscles that move the parts of the tongue, lips, mouth, and throat in order to shape and produce words. Thus, the initial reaction to the input should be symmetrical, because the input is received by both ears if the input is auditory, or both eyes if it is visual. As the task of repeating the word is initiated, this symmetrical pattern on the PET scan should be followed by asymmetrical activity in Broca's area, which lies only on the left side of the brain. Once the word is generated, then the physical task of speaking the word should produce a new, symmetrical pattern of activity as muscles on both sides of the vocal apparatus work to create audible speech. The staggering fact, clearly visible on Raichle's slides, was that translating a word from thought into speech did not provoke asymmetrical brain activity. It is true that a part of the brain near Broca's area was activated in this experiment. Because Broca's area was initially recognized in the nineteenth century, when these sophisticated techniques were not dreamed of, perhaps this part "near Broca's area" can be considered Broca's area itself. The important point is that both this newly defined Broca's area *and a corresponding area on the right side of the brain* lit up together. Another key finding was that this newly defined Broca's area lies deep within the brain. Like a cat lying under the bedclothes, it may make a superficial lump that is visible, but Broca's area itself cannot make a detectable impression on the interior surface of a skull.

This was exciting, but clearly the most important observation for me was the next set of experiments that answered the pressing question: So what does Broca's area *do*? A student was exposed to a word and asked to move his right hand. As expected, this task used the same input pathways; unexpectedly, it then lit up the newly expanded Broca's area. Since the left side of the brain controls the right side of the body, only an isolated, right-hand action produced this pattern of brain activation. (Moving the left hand in response to a word caused the right-brain area that corresponds to Broca's area to light up.) In other words, while the new Broca's area is active during speech, it is also active during other complex activities that in no way produce language. After conducting these and many similar experiments, Raichle's group had concluded that Broca's area was associated with some higher-level control of motor programming, such as the coordination of complex actions involving the right side of the hand or vocal apparatus, *not* with the production of speech itself. Lesions or injuries to Broca's area itself produce stuttering and other motor problems with speaking, as Paul

Broca had noticed in the nineteenth century, but only a defect to a much larger area causes aphasia, or loss of language.

This lecture made a staggering difference in my interpretation of the boy. If the presence of a Broca's area didn't automatically imply that the boy was capable of spoken language, it was time to rethink the whole issue of his communication abilities. Maybe Ann and her colleague Gwen were right. I was energized by this new information. I took a metaphorical deep breath and plunged into the contentious literature on language origins.

As never before, I began to appreciate the importance of language to humans. As long ago as 1863, Thomas Henry Huxley, Darwin's friend and defender, thought that language was the one feature that set the human apart from the animals. He described

> that marvellous endowment of intelligible and rational speech, whereby . . . he has slowly accumulated and organised the experience which is almost wholly lost with the cessation of every individual life in other animals; so that, now, he stands raised upon it as on a mountain top, far above the level of his humbler fellows, and transfigured from his grosser nature by reflecting, here and there, a ray from the infinite source of truth.

Many others since Huxley have located human uniqueness in our command of language; to be fair, there has been a strong lobby opposing this view, and championing the ability of other animals to communicate. The resolution of this debate lies in the very nature of language.

In trying to understand what language is and how it functions, scholars have turned to three main sources. They have studied language capabilities among apes in order to deduce the minimal template for language: the basic ability that was presumably shared by the last common ancestors of apes and humans. They have also documented language acquisition—how infants learn language or how adults learn a new language—in an attempt to discover the underlying and unconscious rules by which language is encoded in our brains. Finally, some researchers have focused on language difficulties or defects among those with impaired abilities, due to congenital or hereditary conditions or to injury. Knowing that an injury to a given area produces a given sort of language difficulty yields telling clues to the language functions of the brain. The results of defects can be extraordinarily specific, such as tiny strokes that cause the inability to retrieve the names of fruits or vegetables, or disturbingly general, such as the fluent but nonsensical "word salad" produced by sufferers from some disorders.

In order to think productively about the origins of language, I had to resolve to my own satisfaction the question of how to define language. Spoken languages clearly dominate among humans; the very word we use, *language*, is derived from the same root as the French word for *tongue,* a term we also use sometimes to refer to a language. What a bizarre twist of evolution it was that a primate like ourselves evolved a reliance on *spoken* language. The zoological order known formally as the primates comprises monkeys, apes, and primitive prosimians like those I studied for my Ph.D. thesis: the galagos or bush babies, lemurs, and tarsiers. Primates are overwhelmingly manipulative animals—they are always handling, altering, and using objects with their hands—and, as a group, they are not especially verbal. If you had to guess at a modality in which primates would evolve an elaborate communication system, you would probably bet on a gestural modality. Nonetheless, humans have followed an evolutionary trajectory that has led primarily to spoken, not signed, language.

The primacy of spoken language among humans is extraordinary. During the days of world exploration by Europeans, no people was ever encountered that lacked language. Indeed, this concept of perpetual speechlessness, of an inability to express language, is so alien that we have no common adjective to describe the condition. We have words with which to talk about written language and its possession; people are either literate or, if they lack written language, illiterate or preliterate. So, by extension, people with language might be described as *linguate* and those without spoken language must be something like *illinguate,* except there are no such words. In 1868, when Ernst Haeckel conjured up his hypothetical ape-man, *Pithecanthropus*—the archetypal missing link who was later embodied as *Homo erectus*—he gave it the trivial name *alalus,* meaning "speechless" or "without speech." We have no English equivalent. *Mute* does not suffice, of course, for it refers to individuals who cannot speak yet belong to a society and a population in which speech is a normal attribute; this term describes an individual's unusual condition rather than an attribute typical of a population or species. I find the lack of words to describe language capabilities fascinating. Other attributes shared by all normal humans have descriptors: we are bipeds; we are brainy or smart; we are relatively hairless or naked; we are upright; we are hearing; we are sighted; we are social. But there is no word in English to describe the ability to communicate through language; of all these important attributes, only language ability is so deeply embedded in our humanness that it is never remarked upon.

And yet it is not the ability to speak that makes us human. Language must not be confused with speech, for many people who do not or cannot speak are still linguate, communicating through sign languages. Of course, the sign languages used by the deaf or hearing impaired are full or true languages. They are often largely or wholly independent of the dominant spoken language of the region; thus, for example, American Sign Language is not a manual translation of English. Its syntax, grammar, and vocabulary do not coincide with that of English.

American Sign Language is simply a language unto itself, used by nonhearing (or hearing-impaired) people who function in a dominantly hearing and speaking world.

There is another, telling proof that language is not the same thing as speech. In a recent study of deaf children born into signing families, Laura Petitto and her colleagues at McGill University showed that, even as normal babies begin to speak by babbling, deaf babies begin to "speak" by babbling *with their hands*. These babies repeated signs or partial signs—the manual equivalent of nonsense syllables like "la-la-la-la" or "mum-mum-mum"—over and over again. As a hearing baby will, these deaf babies tried to join in the conversation by making utterances (signing nonsense words or syllables) at the appropriate points, enacting the rhythm of a dialogue before they have mastered its content. As Petitto observed, the fact that deaf infants babble shows that language is an innate capacity in humans; it is the mode of expression, not the ability itself, that is learned. Or, to use the felicitous phrase of Steven Pinker, a linguist at MIT, there is a "language instinct" hard-wired into the human brain.

Learning a language is a tricky business for a child. There appears to be a critical period during the child's development in which he or she learns how to express language, a transient window of opportunity that, once missed, cannot be regained. Children who are deprived of human contact during this crucial period are not able to learn full language later. A well-documented example is the child known as Genie, who from eighteen months until thirteen years of age was imprisoned alone in a bedroom by her father. Once she entered the outside world in 1970, she was found to have normal intelligence and received intensive instruction in language. Sadly, she never acquired full language. Like other such children, Genie acquired a limited vocabulary (and responded to a much broader one) and simple patterns of grammar and syntax, but she never mastered language. Her usual utterances were remarks such as "Want milk" or "Paint picture." Her most complex utterances were sentences like "I want Curtiss play piano."

The record of Genie's speech demonstrates some crucial distinctions between verbal utterances and full language. Language is much more than a capability to learn and express the abstract symbols for entities or concepts that we call "words." Language is a system of communication, one that implies the existence of at least two individuals who share a common set of conventions and symbols; it is an inherently social activity. This is why someone like Genie, who was socially deprived but not intellectually impaired, did not learn language. At the age at which most children are learning language, Genie had no one to talk to, or to talk to her. These appalling circumstances caused a dreadful stunting of her humanity from which she could never recover fully. Not only the general ability to communicate through language but also the mechanics of any particular language—the specific words or symbols used to refer to an entity (such as *cat* or *gato* or *paka* in English, Spanish, and Kiswahili, respectively)—are transmitted culturally. Without culture, without society, there can be no language.

Language is also fundamentally both symbolic and arbitrary, because words are symbols that have no consistent or overt relationship to the item to which they refer. Language has what Sue Savage-Rumbaugh, a well-known researcher in ape language, calls displaced referents; that is, words can and do act as symbols referring to subjects that are not actually present. Thus, we can talk about clouds whether or not there are any visible in the sky; we can even talk about referents that have never existed in the real world, such as blue dragons. Because of this attribute, language is also productive; new words can be created to refer to new ideas or experiences, the meanings of words may alter over time, and words may even acquire two or more meanings in ways that allow for joking, punning, and other complex wordplays.

Genie's speech has most of these attributes, and here her command of language goes well beyond that of any animal that has been observed. Animal language works through rather limited vocabularies of calls, postures, and sometimes scents that appear to convey concrete meanings: "Danger! Leopard" or "I am sexually receptive." Konrad Lorenz, the great ethologist, has paraphrased the most universal animal signal as, "I am here; where are you?" Animals clearly remember the past and sometimes plot elaborately to manipulate the behavior of others in their social group. Monkeys have been observed to give an alarm call, indicating that a predator is near when it is not, in order to distract other monkeys from a favorite food source. Still, nonhuman animals apparently cannot discuss the distant past, the remote future, or abstract or hypothetical ideas.

True or full language must also include two specific categories of words, according to linguist Derek Bickerton, whose book, *Language and Species,* was one of the most provocative I have read. First there are those words that refer to concrete objects (nouns like *table, toy,* or *dog*), perceptible attributes (adjectives like *green* or *noisy*), and real actions (verbs like *run, hug,* or *give*)—what linguists call lexical items. This category can be extended to include words that express constructs or abstract ideas (*loneliness* or *absence*). Genie, and at least some animals, clearly use lexical items in their language. In additional, true or full language includes a number of words or partial words that are primarily relational (*in, of,* the *'s* that indicates possession), numerical (*any, many, some*), referential (*that, a, this*), temporal (*before, until,* or endings that indicate tense such as *-ed* or *-ing*), directional (*to, at, from*), and so on—which linguists call grammatical items. It is the grammatical items that allow us to express complex thoughts in a single sentence without confusing our listeners; they eliminate ambiguities or, as linguists say, they disambiguate our utterances. These grammatical items transform a sentence like Genie's "Applesauce buy store" into one of its several possible meanings: "I want some applesauce; let's buy it at the store" or "Applesauce is what we always buy at the store" (a more familiar phrasing would be "We always buy applesauce at the store") or "Applesauce is what we bought at the store." Less obvious interpretations of Genie's sentence might include the meaning that a person named Applesauce bought a store or the statement that the food, applesauce, could be used to buy a store in an apple-loving society. The grammatical items that are missing from Genie's remarks are decoders of meaning and unscramblers of reference. In fact, one of the awkward linguistic habits of American scientists is to omit many of these grammatical items from their writings, for which a substantial cost is paid in terms of clarity.

In contrast to full language, small children, individuals like Genie who have missed the opportunity to learn language normally, and apes who have undergone considerable training all use a much simplified form of language. There is usually only one tense, the present tense. The structure of the utterances or sentences is very simple—"Me up" or "Give me banana me banana me"—and contains few or no clauses. In these circumstances, part of the "sentence" is often gestural: "Open" accompanied by a fixed look at a cupboard and then a fixed look at the trainer (to convey "I want you to open this cupboard for me"). In fact, my cat is adept at conveying exactly the same message without bothering to verbalize. More to the point, grammatical items are rudimentary or, often, completely absent.

This restricted or bare-bones language is what Bickerton calls proto-language. He believes it is the first means of verbal communication that we learn as children and is probably a fair approximation of the first means of verbal communication that we developed evolutionarily too. It is the form of language that we share with a few talented and trained apes. Bickerton suggests that proto-language is a robust if limited means of communication that survives even horrendous deprivation like that meted out to Genie or those who suffer particular types of neurological injuries. It is the fallback rudimentary type of language also used by people fully adept in one language who are trying to communicate in another; thus, proto-language lies at the root of pidgin languages. Proto-language is the sort of language we can readily envision as developing by small increments from the extant oral and gestural utterances of many social species.

Bickerton argues that proto-language and true or full language are two systems separated not only by their modes of expression but also by their evolutionary genesis. In his view, proto-language and true language evolved independently to serve different purposes, and they probably have different neurological bases. This is why proto-language does not become full language as the speaker matures or learns more. Genie (or a trained ape) does not suffer from arrested development of language; she has fully developed proto-language and has failed entirely to develop the other system that is true language. Under normal conditions, proto-language is supplemented and eventually supplanted by full language in humans.

Why have apes failed to learn full language? It is not because they are physically ill adapted for speech (which they are) nor is it because they cannot grasp the use of symbols. Experiments conducted by Allen and Beatrice Gardner, working with the chimpanzee called Washoe, by Penny Patterson with the gorilla Koko, and by Sue Savage-Rumbaugh with the pygmy chimp named Kanzi have all demonstrated that apes have an impressive ability to learn symbols and icons. Because her protocol involves a computerized board with lexigrams or arbitrary symbols on it, Savage-Rumbaugh's work with Kanzi has effectively demolished the criticism that ape language was a product of wishful thinking on the researchers' parts. Clearly apes exposed to appropriate linguistic opportunities learn to combine symbols into multiword utterances and to participate in meaningful dialogues. The problem, according to Bickerton, is that apes do not have the elaborate representational system that humans possess and so they never progress from proto-language to full language. There is an absolute limit to the complexity of their utterances, a limit that is both grammatical and conceptual.

Bickerton hypothesizes that proto-language developed as a communication system, based on the neurological template that we share with apes. In contrast, he believes that the neurological basis for full language evolved as a complex system for taking in sensory information about the environment, processing it, storing it, and perhaps evaluating it as a basis for future actions. The basis for full language, argues Bickerton, was a sort of mapping function, a means of representing the world internally. While all creatures map their world to some extent—trout have exquisitely accurate templates of the shape, size, and behavior of suitable prey, which is what makes fly fishing so challenging—humans have evolved a stunningly intricate representational system that far exceeds that of other organisms in complexity and subtlety. The more complex this internal or mental representational system is—the more categories we can create for classifying the infinite number of items, sensations, and actions that we sense or think about—then the more distanced we are from the reality before us. In other words, in order to make a highly detailed and accurate map, one that changes minute by minute as new information is added, we interpose a tremendous amount of mental processing between the experience and our mental representation of it. This distance has the advantage of freeing us from the tyranny of the present: it permits us to think about circumstances or events that are not occurring and may never occur. And, Bickerton notes, in order to achieve consciousness—in order to "stand outside yourself" and look at (or think about) yourself—there must be somewhere else to stand. Without a detailed mental symbol that represents yourself, you cannot think about yourself in any complex way. Apes seem to have only a rudimentary sense of self and a limited degree of consciousness, but they simply lack the elaborate representational system that would enable them to develop truly complex thoughts and full language.

What did this mean for the Nariokotome boy? His vertebral anatomy suggested to Ann that he had no facility for verbal language; it rendered him speechless, in fact. But was Ann's suggestion correct? Would other analyses confirm it? And, if they did, how would it change my image of this boy I thought I knew so well?

14

FINDING LINKS,
MISSING LINKS

My foray into the literature on language origins gave me several specific ideas to think about. Language is predominantly spoken, meaning that the anatomical capacity for speech has to exist, but it also reflects particular mental abilities. These include the ability to map, categorize, and analyze the world in a complex fashion. Language also requires the use and understanding of symbols (or displaced referents), the capacity to create novel symbols (i.e., to be linguistically productive), and the habitual practice of using both lexical and grammatical symbols. Language capability is developed through social interactions, and the vocabulary, grammar, and syntax of any particular language are culturally transmitted. Derek Bickerton's work, in particular, had given me the idea that language might evolve in two stages, through two separate routes. The challenge now was to detect the origin of these attributes in the fossil record.

Anatomical capability was one topic I knew something about. Now that Broca's area was effectively eliminated as ironclad evidence of speech, what was left? For some years, several researchers had been trying to establish the shape and size of the vocal apparatus in different types of early hominids. Because the vocal tract itself is composed of soft tissues, which do not fossilize, the only clues from fossils are the hyoid and the subtle markings and shapes of the base of the skull, which can be ambiguous. The hyoid is a small and fragile bone that anchors the tongue muscles; it is rarely preserved and is unknown in the fossil hominid record until Neandertals appear about 1.5 million years after 15K lived. This leaves only the anatomy of the base of the skull as the basis for reconstructions of the vocal tract.

The first, groundbreaking attempt at such work was started by Phillip Lieberman and Edmund Crelin in 1971, with a project aimed at determining the speech capacities of Neandertals. Lieberman, a speech analyst, and Crelin, a gross anatomist, reconstructed the vocal tract of a particular specimen, known as the Old Man of La Chapelle-aux-Saints. They worked from a cast of the skull that was commercially available, but they were unfortunately unaware that the base of the skull had been broken away and inexpertly reconstructed before the casts were made. However excellent a cast of a fossil is, I know that the details always must be checked, to ensure that the original is faithfully portrayed. But neither Lieberman nor Crelin is a paleoanthropologist and they were oblivious to this pitfall. Their idea was a good one, but the anatomy they used as their starting point was open to serious dispute and their work was severely criticized in the anthropological literature. Their conclusion, that Neandertals may have talked but would have had a greatly restricted range of vowel sounds, was not generally accepted.

But an anatomist/anthropologist of the next generation, Jeff Laitman, persisted, introducing new methods and bringing greater rigor into the attempt to reconstruct ancient vocal tracts. He developed measurements that showed the relationship between the increasing flexion of the base of the skull and the development of a more and more human vocal tract, with a larynx placed low in the neck. Laitman showed that, as hominids evolved from apelike forms to australopithecines, *Homo habilis, Homo erectus,* Neandertals, and finally modern humans, cranial flexion went from nonexistent in australopithecines and *habilis* to substantial in *erectus.* Cranial flexion was full blown in Neandertals some 300,000 years ago. A series of careful analyses convinced Laitman that the earliest hominids, like the australopithecines and habilines, were anatomically unable to talk; like me, he had read Raichle's work and knew that having a Broca's area did not mean that these species had language. But Laitman's studies left the case for language in *erectus* equivocal, for 15K's larynx probably rested in a position very similar to that of a young modern human child. Toddlers speak proto-language, so the boy's ability to vocalize might have been developed to a similar degree.

If *erectus* was anatomically capable of some kind of speech, did that make the boy and his colleagues linguate? Possibly, but confirming evidence of the other attributes of language were needed before I could endorse such a conclusion. The next criterion would be evidence of the regular use of symbols or icons. When I started to examine the record of human behavior—the archaeological remains left behind by different hominids— searching for clear symbols, I was walking on well-trodden intellectual ground. Previous attempts to identify symbolic behaviors in the archaeological record had produced two widely divergent answers.

Historically, the first answer was that the manufacture of stone tools shared many of the features of the cognitive processes needed for language. Productivity was shown by the variable forms of artifacts, which are manufactured by combining a basic "vocabulary" of motor operations in the same way that words are combined into phrases or sentences. Arbitrariness or symbolic content was seen in the imposition of a predetermined form upon a raw material. In other words, hominids seemingly had in their minds the standardized shape of a particular tool, such as a teardrop-shaped hand ax, and altered lumps of rock until they conformed to this shape. Cultural transmission of symbols was read into the repeated creation of a similar set or industry of tools by the same evolving population over time, but other aspects of language could not be detected or could not be analogized with stone tool manufacture. If making stone tools in regular shapes does correspond analytically to the processes used in creating language, then the origins of language may go back to a period about 1.4 million years ago. That is the point at which some hominid—it is generally believed to be *Homo erectus,* but no one knows for sure—began making hand axes and the consistent flakes that are struck off of them. Before 1.4 million years ago, stone tools were made, but their shapes are inconsistent and highly variable.

The "language written in stone" idea was first expressed by Ralph Holloway of Columbia University in the late 1960s and has been elaborated upon or challenged many times since. Research subsequent to Holloway's initial publication has revealed a new view of stone tool manufacture that suggests flaws in this analogy. It is now clear that the form of an artifact, and the sequence of motor operations needed to produce it, are largely dictated by the raw material itself. Instead of a hominid's "deciding" to make a hand ax, for example, we now understand that it is the size, shape, and fracture properties of the chunk of raw material at hand that determine whether or not a hand ax will be made. This means that the location of a toolmaker on the landscape relative to sources of various raw materials may be the overriding determinant of the components of that hominid's tool kit. The persistence of a characteristic suite of tools in one area over time is not necessarily evidence of cultural transmission of tool types and technological skills, as was long believed; the tool kit may simply reflect what can be done with the quantities and types of stone that are found in that region. Some tools can be made of large pieces of stone, others of small; some tools must be made of fine-grained rock, others are less demanding; and so on. Technological breakthroughs—inventing or learning new ways of shaping stone—may increase the options of what can be made with a given raw material, but the limitations are imposed by the material, not the toolmaker.

Another problem that came to light after Holloway made his initial suggestion can be stated as a simple question: If toolmaking indicates the possession of the cognitive faculties necessary for language, how is it that chimps can and do make and use tools (in the wild and in captivity) and yet never master full language, even with intensive training?

Other analyses have suggested that language was a very late acquisition indeed; this made more sense to me from my new perspective. Two anthropologists at the University of New England in Australia, William Noble and Iain Davidson, have emphasized the symbolic nature of language as its most readily visible and perhaps the most important attribute. To them—to me too—the earliest, unequivocal evidence for the repeated use of symbols occurs very late. This evidence occurs at different times in different parts of the world, perhaps reflecting the spread of modern humans. Nowhere do undoubted symbols appear earlier than about 125,000 years ago, when anatomically modern humans first evolved; in much of the world, symbolism appears a mere 30,000–50,000 years ago. (While some people think of 50,000 years ago as very ancient, to me it is a negligible span of time that is smaller than the dating error of the fossils I work on.) Noble and Davidson point to abundant evidence from the Upper Paleolithic, a period that began around 35,000 years ago in Europe coincident with the appearance of anatomically modern humans. Although there are a few cases of apparently symbolic behavior in Europe earlier than this, they are isolated instances and thus not fully convincing. But unmistakably symbolic signs occur, repeatedly, from the beginning of the Upper Paleolithic onward.

Among the earliest known symbolic expressions are the beautiful little stone carvings from Vogelherd, Germany, a site dated to about 32,000 years ago that yielded caches of carved horses, humans, and other figures associated with arbitrary signs. Similar symbolic and arbitrary expressions abound in the Dordogne region of southwestern France, in the spectacular painted caves such as Lascaux or in the less well known but equally stunning rock shelters filled with bas-relief sculpture, elaborately carved bone and ivory objects, and engravings. The animal depictions are amazingly beautiful and powerful, as well as being readily recognizable. Less well known (but not less common) are the geometric or linear symbols—odd grids, repeatedly used angular constructs, rows of dots, zigzags, V-shapes, and the like—that accompany the animals. The meaning of these geometric figures (or, for that matter, the meaning of the animal representations to the original artist) is so opaque as to defy analysis. Some items have been suggested to be arrows or traps, others to be "female symbols," still others to be artists' signatures. Some may convey no more than "I am here."

Yet without doubt all of these images are symbols and all of them are meaningful, even if we are today uncertain of the message. They were made with great care and considerable effort on the part of the artists. The sites where they were created are not always easily accessible; the pigments had to be gathered, ground, and carefully prepared; scaffoldings and lights were needed in many locales (and their traces have been found). Huge bas-relief carvings, as in the rock shelter known as Cap Blanc, or elaborate paintings, as in Lascaux, probably took days or even weeks of effort. The creation of the paintings, sculptures, and engravings seems likely to have been a ceremonial occasion, perhaps accompanied by music or song, according to studies of cave acoustics and the placement of artworks. These artistic expressions were much more than idle doodling or scribbling; they were deeply important to the artists, and I believe they are deeply important still. Almost certainly these are not the earliest symbolic (and hence linguistic) behaviors in the human lineage; they are simply the convincing earliest *evidence* of manifestly symbolic behaviors. In evolution, new behaviors routinely precede the appearance of concrete adaptations that facilitate those behaviors.

There are hints that symbolic behavior may have occurred earlier than the origin of modern humans. For example, our predecessors in Europe, the Neandertals, buried their dead; we find whole skeletons in deliberately dug trenches, sometimes associated with lumps of red ocher, or a tool or two, or a segment of an animal's leg. The dead are sometimes arranged in artificial postures known as flexed or crouch burials, with the knees drawn up tightly; bodies were certainly manipulated and may have been bound with some sort of fiber to achieve this position. Some archaeologists have argued, perhaps too fervently, that this mortuary ritual is clear evidence of a spiritual belief in an afterlife for which the dead person had to be prepared and equipped. While that interpretation leans too heavily on Western religious beliefs to persuade me, I think these burials are at the least evidence of some careful housekeeping and respectful treatment of individuals who are no longer alive. Is burial of the dead a symbolic act? The Upper Paleolithic behavior leaves no room for doubt, for the most spectacular of Upper Paleolithic burials have all the attributes of Neandertal burials and much more: intense patches of red ocher that must have been scattered over or painted onto the bodies; caps or cloaks covered in beads made of animal teeth or shells; carved bracelets, pendants, and other personal adornments; tools; and in one case, two pairs of mammoth tusks. These elaborate grave goods and body treatments unquestionably reflect symbolic behaviors. The difference between Neandertal and Upper Paleolithic burials is the difference between a burgeoning ability and one that is so fully developed as to be unmistakable. It may also be the difference between proto-language and true language.

If this interpretation is correct, then the advantages of more sophisticated and precise communication with others may be exemplified in the results of studies conducted by Olga Soffer, an archaeologist at the University of Illinois at Urbana-Champaign. She has looked at the differences in settlement patterns between earlier, Middle Paleolithic sites of northern Eurasia, presumably made by Neandertals, and those of later sites of the Upper Paleolithic in the same region, presumably made by modern humans. Earlier sites are smaller in area, suggesting people lived in smaller social units, and the sites' contents suggest their residents used strictly local resources. Earlier sites are also geographically or ecologically restricted, occurring and recurring in the same areas. In contrast, Upper Paleolithic sites are very different: they are larger, they contain items derived from more widely scattered resources, and they are located in more diverse habitats. Skeletal remains from these periods reflect differences between Middle and Upper Paleolithic lives. The Neandertals from the earlier sites were subjected to more physical stress and died younger than anatomically modern humans, who survived weaning and childhood in greater numbers and enjoyed better health. Soffer concludes that the transition reflects "a dramatic change in economic and social relationships" that coincided with the appearance of anatomically modern humans. To me, this change appears to be the aftermath of the development of true language, with its consequent improvement in planning and in the sharing of information.

Another way of looking at the origin of language is to ask what language is *for*. It is, according to Bickerton, a sophisticated system for representing the world that has secondarily been usurped for communication. Minimally, language, even proto-language, implies the existence of two individuals, a speaker and a listener. It is *about* social interaction and the exchange of information, which implies that the speaker and the listener do not share all knowledge in common. We might ask, then, when it becomes clear in the fossil record that groups of hominids began encountering other groups of hominids who were sufficiently foreign to make the exchange of information an important adaptive mechanism. Once again, the answer points to the Upper Paleolithic. This was the period when objects of personal adornment first began to appear regularly. As Randall White of New York University has observed, clothing, jewelry, and makeup or body paint are all means of projecting an identity. Then, as now, personal adornments almost certainly symbolized both individual identity and that individual's allegiance or membership in some larger group. What clothes or jewelry you wore, your style of body paint or scarification, or your haircut was a symbolic way of telling others who you were. There was no need for personal adornment as long as everyone was familiar; as in a small village, everyone would know everyone else from birth. The lack of personal adornments attests to a small social world, restricted perhaps to a few wandering bands who encountered one another regularly. The rise of personal adornments in the Upper Paleolithic, the greater density and larger size of archaeological sites, and the contents of those sites show clearly that people traveled substantial distances to congregate (at least periodically) into much larger groups. This new pattern of living meant that suddenly there was both a need and an occasion for demonstrating visually that you were part of this group and distinct from that. Lines were drawn between us and them; ethnicity was born.

It is easy to envision that these periodic gatherings would also have been occasion for important exchanges: of potential mates and of information. Margaret Conkey, a Berkeley anthropologist, has argued that the symbolic art of the Upper Paleolithic was a means of encoding information. Art was in some sense an aide-mémoire, needed because new, richer, and more diverse information was now available through communication with others. Art and ceremonies not only created images, they created memorable occasions, experiences that evoked strong emotions that would embed the information firmly in the mind of the participants.

We can only imagine what was communicated: places to hunt; new sorts of traps; locations of water, good caves, or outcrops of stone good for making tools; the location of plants or herbs that might be edible or might heal illnesses; techniques for making tools, traps, or snares; information about the behavior of animals; tales about the weather and climate; or ways to make and keep fire. Mixed in with all of these topics, and probably others we cannot imagine, would surely have been those most human of all interests, personal gossip and stories. Knowledge of places, resources, events, and people would be valuable and precious information. Symbols were one means of remembering this information, of mapping the world permanently.

There are other sorts of evidence that suggest advanced cognitive skills on the part of early modern humans. The ability to colonize new continents, especially Australia, also bespeaks considerable cognitive sophistication on the part of our ancestors, even if there are no archaeological remains that directly demonstrate arbitrary, symbolic behavior. Greater Australia (a region that includes the continent of Australia and several adjacent islands that were then part of the same landmass) was colonized at least 40,000 years ago and maybe as many as 60,000 years ago, according to the oldest known skeletal evidence and archaeological sites. Archaeologist Sandra Bowdler of the University of

Western Australia sees the colonization of Greater Australia as the culmination of a series of waves of territorial expansion that brought significant numbers of humans into southeast Asia for the first time since *Homo erectus*. In any case, this island continent could not have been reached from the mainland without boats; to reach Greater Australia required a trip across some 200 kilometers (125 miles) of open sea, heading toward a landmass that could not be seen from the shore. The trip must have been repeated many times, for the archaeological evidence shows that the continent was populated too rapidly for the colonizers to have been a single boatload of people and their offspring.

The difficulty of colonizing Greater Australia was not over once the first colonists had survived the trip. Bowdler believes that the colonizers were tropical rain forest peoples with a strong set of adaptations for the exploitation of coastal or aquatic resources. They may have followed the shoreline and rivers down the eastern shore of Australia, where rain forests persisted. Yet many things about the new continent were different and difficult. None of the animals the colonizers knew and hunted were there, and indigenous Australian species took the place of many of the familiar rain forest plants. Many aspects of their previous adaptations to life needed to be altered, particularly as they came to drier areas or highland regions of Australia. Most of the desert interior of Australia was simply unsuitable for human habitation, as many have discovered at their peril. Considering the complexity of the problems that had to be solved to build suitable ships, accomplish such voyages, and survive, reproduce, and populate a new continent—problems that caused high death rates among English immigrants and deportees who tried to recolonize Australia in the nineteenth century—I find it hard to imagine that the people who first colonized Australia lacked full language.

Inspecting the archaeological record is not the only means of investigating the origins of human language. An entirely different type of evidence was gathered by Luigi Cavalli-Sforza, a prominent human geneticist at Stanford University, and his Italian colleagues. They collected blood samples from forty-two different human populations around the world: Mbuti Pygmies, Lapps, North African Berbers, Sardinians, Eskimos, Melanesians, Indian groups from North and South America, Europeans, Tibetans, Maoris, and many others. The team analyzed these samples for information about the distribution of 120 alleles, or genetic alternatives. Using complicated statistical techniques, they grouped the results into phylogenetic trees that reflected the genetic resemblances among the human populations. These trees provide not only an estimate of relationships but also a sequence of branching that can be assumed to represent the path of evolution.

Their results led to some striking conclusions. Reassuringly, the genetic data from different populations cluster into groups that correspond to geographic realities: Africans with Africans; Asians with Asians; Australians with other Pacific peoples; and so on. Analysis of these clusters and the distances among them supports the idea that all modern humans shared an African origin. The greatest genetic distance lies between the African populations and all other groups combined, implying that the Africans separated first from the other groups, giving Africans the longest time over which to evolve independently. Cavalli-Sforza concludes that the first split occurred when the initial, single population divided into an African and an Asian group; then Australians broke away from the Asians; and finally the Europeans separated from the Asians. This branching pattern would fit well with a model of migratory waves of modern humans spreading across the Old World, starting from an African origin. Cavalli-Sforza sees a congruence between the genetic distance data and the fossil evidence for the spread of anatomically modern humans into various parts of the world. Modern human remains are about 125,000 years old in Africa, about

50,000 to 60,000 years old in Asia, about 40,000 years old in Australia, about 35,000 years old in western Europe, and only 15,000 years old (with contested evidence up to about 35,000 years) in the Americas.

Cavalli-Sforza's argument becomes compelling when he compares his branching phylogenetic tree, based on genetic data, with one produced by a group of linguists trying to reconstruct the "evolution" of modern languages. The two trees—one linguistic, one genetic—are amazingly similar, so similar that the two seem highly likely to reflect the same historical events. Of course, the language you speak is not genetically determined but culturally transmitted. There are many well-documented instances of peoples who have lost their original language and adopted that of a socially dominant group. In fact, a language is more easily replaced than (in most cases) genes are; languages change easily while it seems relatively rare that the genetic makeup of one population is swamped by an influx of new genes from another. Thus Hungarians speak a Uralic language originally imposed on them by the Magyars who conquered Hungary in the Middle Ages; nonetheless, genetically, Hungarians are predominantly European, with only slight traces of Magyar genes.

Genes, though, can move surprisingly quickly. Cavalli-Sforza gives an example closer to home for many of us: African Americans today have a gene pool that is on average about 30 percent European, yet African slaves and their descendants have been in America for no more than two hundred years (and often less). One way in which this admixture could occur in such a short span of time, Cavalli-Sforza calculates, is if 5 percent of all children born to African Americans in each generation (since the institution of slavery) had one European and one African parent and if all of those offspring were considered to be ethnically black.

Although languages evolve more rapidly than genes and move more freely among peoples, there are only about five thousand languages in the world today. Of course, the number of individual languages is only a general reflection of the antiquity of language as a capability; you can't calculate the time of the origin of all languages from such information. But surely if humans had had language since the origin of the genus *Homo,* back almost two and a half million years ago, the number of languages would be far greater and the diversity of tongues broader.

Cavalli-Sforza makes one final, telling observation about the link between anatomically modern humans and full language. Neandertals disappeared and were rapidly replaced by modern humans, an event that may have taken only a few thousand years in some parts of Europe. He finds the strong dominance of modern humans easier to understand if Neandertals were not fully linguate, "if they were biologically provided with speech of more modest quality than modern humans," as he puts it.

> In our society [Cavalli-Sforza writes], until 100–150 years ago, deaf-mute people had very little chance of reproducing because of strong adverse social selection. . . . Even if interfertility was potentially complete and there was little or no impingement, Neanderthals must have been at a substantial disadvantage at both the between- and the within-population level.

In my mind, these varied lines of evidence—anatomy, archaeology, and genetics—all point to a single conclusion. True language seems to me to have been a very recent acquisition, one that just precedes and enables the evolution of anatomically modern humans and fully modern behaviors. It would seem that, once again, Haeckel's unfounded guess about the attributes of the missing link was correct: *Homo erectus* was speechless, illinguate. Not only does this conclusion contradict the accepted wisdom that language acquisition demarcates the origin of the genus *Homo,* it leaves me with a haunting and novel image of the Nariokotome boy.

Here was a young man, tall, black, lean, and muscled, thoroughly adapted to his environment. He made tools that, although crude, represented a substantial advance over those of his predecessors and he made these tools according to a distinct and repetitive plan, using deliberate techniques. He lived in a group with strong social ties, one that nurtured helpless infants and nourished their mothers. He and his kind were very successful in obtaining high-quality foods, almost certainly by hunting, so successful that the evolution of big brains and large bodies could occur. The boy's species, *Homo erectus,* was perhaps the cleverest that had yet walked the face of the earth. Long-legged and immensely strong, this species strode out of Africa. They were such effective predators that they could invade and colonize most of the Old World at a rate that appears virtually instantaneous to our modern dating techniques: less than a hundred thousand years to get from Africa to Java, not by deliberate migration but by simple population expansion, year after year.

All of this looks and sounds so human, and yet . . . and yet the boy could not talk and he could not think as we do. For all of his human physique and physiology, the boy was still an animal—a clever one, a large one, a successful one—but an animal nonetheless.

This final discovery of the boy's speechlessness had an enormous emotional impact on me. Over the years that had passed since Richard, Kamoya, and I had first excavated his bones, I had thought I was growing to know the boy, to understand him, to speak his language, metaphorically. I grew fond of his form; his face took on the familiarity of a member of the family or an old friend. I could almost see him moving around the harshly beautiful Turkana landscape, at a distance looking enough like the Turkana people to be mistaken for human. *He did this,* I would think, *he knelt there to scoop up water or crouched behind a bush like this one to stalk an antelope.* But then, as I approached him closely, preparing mentally to hail him and at last make his acquaintance in person, it was as if he turned and looked at me. In his eyes was not the expectant reserve of a stranger but that deadly unknowing I have seen in a lion's blank yellow eyes. He may have been our ancestor, but there was no human consciousness within that human body. He was not one of us.

EPILOGUE

Had we found the missing link?

Oh, yes. *Homo erectus* always has been and always will be *the* missing link. Dubois knew it was in 1892, when he realized the fossils he had found could be fitted into Ernst Haeckel's speculative hominid phylogeny. I can think of no better historical precedent for identifying the missing link than that.

To be accurate, of course, it was Kamoya who found the missing link, the Nariokotome boy we nicknamed 15K after his museum number. But it was I who, with many colleagues, deduced the meaning behind the boy's skeleton. This team of creative, hardworking, and sometimes iconoclastic scholars wrung amazing insights out of that singular collection of bones and found out more than I ever anticipated, certainly more than anyone ever had before about *Homo erectus*.

What I learned from the missing link was surprising and un-nerving. My colleagues and I came to the specimen with a legacy of decades of emphasis on the importance and extraordinary size of the human brain: from the arrogance of calling ourselves *Homo sapiens,* to Keith's cerebral Rubicon for inclusion in the Hominidae, to Louis Leakey's conviction that the toolmaker at Olduvai had to be the big-brained *Homo habilis* and not *Australopithecus boisei,* to the belief that Broca's area on the endocast of 1470 marked the origin of language. Even though I rejected many of the specific claims put forward by this scientist or that for the primacy of the human brain and its fundamental role in hominid evolution, I was not immune to a subtle shading of my expectations. Somehow, without my knowing it, I had been in-fluenced by this legacy so that I shared an unconscious expecta-tion that the missing link would have certain characteristics— meaning that evolution in our lineage had followed a particular pathway.

When we discovered 15K, there was a great divide between apes—animals in animal bodies—and humans in human bod-ies. This much was obvious. What was less clear was that the long history of brain-first theories in paleoanthropology implied that the missing link lying between apes and humans would be behaviorally and intellectually human while trapped in an ani-mal body. In that case, the fossil and archaeological record would yield evidence of human capabilities and behaviors while the bones themselves would speak of apelike proportions and imperfect physical adaptations to human behaviors.

But the Nariokotome boy proved the inverse. He was in many ways an animal in a human body. Over and over, when we examined his remains, we found humanlike attributes: an anatomy of legs, pelvis, and torso that was nearly modern in its adaptations to upright posture and gait; a vestibular apparatus similarly adapted to balancing on two legs; the human propor-tions of the body; and the unexpectedly large size and massive strength of that body. Even the boy's growth patterns were hauntingly human, from the sequence of eruption of his teeth, to the fusion of his bones, to the way his face would grow to assume adult dimensions. Our analyses even revealed that *Homo erectus* experienced the typically human prolongation of childhood, a pattern that must be closely tied to an extended period of intensive learning. Physiologically, the boy showed human adaptations too. He had a thoroughly human mode of adapting to his tropical climate in terms of body build and heat dissipation. He, unlike so many tropical animals and like humans, remained active during the heat of the day. Even more remarkably, our studies of his pelvis and brain size demonstrated that the missing link had already mastered the human evolutionary trick of bearing big-brained babies and growing their brains to impressive size. It was astonishing that these complex adaptations, attributes that seem so sophisticated and uniquely human, were well in place by 1.5 million years ago.

Of course, there were signs of human behavior too, but these did not paint the picture I expected. The diseased skeleton, 1808, gave heartrending evidence of the development of strong social ties much earlier than I would have expected. Other aspects of the research strengthened my conviction that *Homo erectus* was probably the first hominid to practice systematic, effective hunting, an attribute that I believe came later than most anthropologists would have expected. To me the large brains and large body size of *Homo erectus,* combined with its dramatic geographic expansion, were compelling evidence that, with this species, hunting became efficient enough to ensure regular access to high-protein animal foods. Part of the hunting success came from cooperative efforts, I am sure, but part may have been due simply to intelligence and cunning.

But for me the most telling discovery revolved around language. I had never anticipated that *Homo erectus* would prove illinguate. Not until Ann MacLarnon's analysis of the vertebral

column and the stunning new evidence of brain function derived from PET scans undermined the credibility of language in
early *Homo* did I see how intimately language is a part of being
human. At some deep level, being fully human is predicated
upon being linguate. This meant that the boy my colleagues and
I spent so many years discovering and analyzing was profoundly
in-human. He was large, he was strong, he was a toolmaker, a
hunter, and an intensely social animal adapted to a rigorous, active life in the tropics. But he was not human, did not think like
a human, and could not speak. Had I met him in the flesh, we
would be unable to communicate, not because we had no *language* in common—the problem I experienced in meeting the
Turkana woman described in the prologue—but because we
had no *perception* in common. The truth is that language is as
much a perceptual mechanism, a means of ordering and understanding the world, as it is a communicative device. To me the
boy is a creature of human size and appearance, with super-
human strength and a peculiar combination of tender sensibili-
ties and a practiced, predatory cunning, an almost human whose
perception of the world is utterly alien and incomprehensible.

This image fills my mind with questions I cannot yet answer.
He was an extraordinary combination of the familiar and the
strange, an animal that looked and walked and sweated and
cared as we do, yet one who could not talk and whose ability to
make mental maps of his world was extremely limited. The con-
trasts and inconsistencies are difficult to resolve. How could an
animal the size of a large, strong, tall youth of fifteen manage
with the brain of a toddler? What combination of mute cunning
and amazing strength enabled *Homo erectus* to spread out of
Africa, the continent that had contained its ancestors for millions of years? What happened to the ecosystems it invaded? We
do not yet know.

Surviving with minimal technology in a hostile world is difficult, and hunting is an uncertain means of survival. Both are

complicated if the hunter moves to unfamiliar territory. How did *erectus* plan and kill and keep its young safe in totally novel surroundings? Perhaps the intense social bonds helped; perhaps the pooling of resources and information gave *erectus* an edge. We can only wonder how cooperation and intense bonding were fostered among members of *erectus*'s social groups without true language or how resources were comprehended with their limited mental capacities. Though population densities of *erectus* were undoubtedly low, I also think about what happened when one *erectus* met another as strangers: Did they communicate with each other? How? And what did they say? These may seem to be unanswerable questions—indeed, they *may* be unanswerable questions—but they are worth pondering. You do not recognize answers to questions that have never arisen.

For all the surprises of this endeavor, meeting the missing link in the person of the Nariokotome boy confirmed my expectations in other ways. My long-standing perspective on the whole of human evolution has not always been popular, but I think it is right. The more common perspective is to cast the course of hominid evolution as a trajectory *toward* humanness, as if evolution had a preordained goal. If that were true, then the boy ought to be a perfect missing link, half ape, half human. Instead, he is neither one nor the other but a novel combination of characteristics. To me, it is clear that evolution is a more random process. The various hominid species we know so far are experiments with different combinations of traits from both apes and humans; some of the outcomes have been seemingly haphazard and short-lived, others successful and enduring. But these extinct species were not failed attempts to become human. They were creatures adapted to niches and life-styles that go beyond those of the few species who still survive. One of the great lessons of studying human evolution is that ancient hominids are diverse. There were many ways to be a hominid and, as the record of ape evolution improves, I suspect there will prove to have been many ways to be an ape. Those species that survive today form a paltry shadow of what once was. In an odd way, by showing us his unexpected adaptations and by revealing so much of his life, the boy has resolved the dichotomy between apes and humans by forging a broader definition of "hominid" that encompasses aspects of each.

Now I am anxious to find more fossils, so their skeletons can help us fill in the many blank spaces in our map of hominid evolution. Like the old maps of Africa, the few regions that are known seem minuscule compared to the vast unknown. Colonial surveyors used to label huge tracts of the continent with the initials MMBA for "miles and miles of bloody Africa." Now we might do the same. While a few regions of time and space are well understood, there are still millions and millions of years of "bloody australopithecines" about which we know very little. This one good skeleton of *Homo erectus*, a rather late hominid, has done much to establish the parameters of the hominid adaptation at about 1.5 million years. The skeleton of 15K has shown us how essential and early some attributes are, and how recently others have been incorporated into the human adaptation.

Could we be mistaken, because we have analyzed only one individual? It is possible. But the overwhelming probability, with any fossil, is that it represents an individual who is roughly typical of his or her species. (The exception is, of course, individuals who are grossly pathological, like 1808 with the bone disease that has altered the shape and texture of her bones.) Since the vast majority of individuals are of middling height and weight, the random processes of fossil preservation are likely to give us specimens drawn from that more-or-less average group. The need to check this assumption of normality is the reason it has been so important to compare the information we gleaned from the boy to the other, more fragmentary remains that had been collected earlier. If he had been wildly different from the others (he was not), we would have suspected that we had found an unusual individual and not a typical one. Instead, he seems average, and by virtue of his extraordinary completeness, he has given us the framework within which we can now understand the other material properly. And, as more specimens of *Homo erectus* are found, there will be important and valuable work to be done establishing the physical variability within the species.

But that is someone else's job. I feel that we already have most of the answers about *Homo erectus*, and, like Dubois, I am only interested in the first 90 percent. I am impatient to move on to new problems now, to turn to the very beginning of hominid evolution, which must have been five or six million years ago when the African ape and human lineages diverged. One more good skeleton, a few million years older than the boy, and we can glimpse the origins of the hominid adaptation, examining the biology of apes and humans when they were almost the same thing.

As I write, Kamoya and the hominid gang are out there in the Turkana desert, looking in four-million-year-old beds for another missing link. One day there will be a phone call to Nairobi: "This is Kamoya. We've found another hominid. I think you should come." I cannot wait to see what they find this time.

NOTES

CHAPTER I

PAGE

7 "Mac wasn't high-grading" An account of his life story and the finding of the Nariokotome boy appear in Kamoya Kimeu, "Adventures in the Bone Trade," *Science 86* 7 (1986): 41.

CHAPTER 2

31 " 'It is like confessing' " F. Darwin and A. C. Seward, eds., *More Letters of Charles Darwin* (London: Murray, 1903), vol. 1, p. 41.

32 " 'last link in the chain' " Richard Owen, "Of the Anthropoid Apes and their Relations to Man," *Proceedings of the Royal Institute of Great Britain 1854–1858* 3 (1855): 26–41. A useful discussion of the concept of missing links and the Great Chain of Being appears in T. D. McCown and K. A. R. Kennedy, eds., *Climbing Man's Family Tree: A Collection of Major Writings on Human Phylogeny, 1699 to 1971* (Englewood Cliffs, N.J.: Prentice-Hall, 1972).

32 " 'a sort of ring' " Henrietta Litchfield, ed., *Emma Darwin: A Century of Family Letters, 1792–1896* (London: Murray, 1915), vol. 2, p. 230.

33 *"Pithecanthropus alalus"* Ernst Haeckel, *The History of Creation, or the Development of the Earth and Its Inhabitants by the Action of Natural Causes. A Popular Exposition of the Doctrine of Evolution in General, and That of Darwin, Goethe, and Lamarck in Particular,* trans. E. Ray Lankester (New York: D. Appleton and Co., 1868), 4th American ed., 1909; see especially pp. 363–79, 398–400, 405–6, and 438–39.

34 "Eugène Dubois" The best reference to Eugène Dubois's life and work is Bert Theunissen, *Eugène Dubois and the Ape-Man from Java: The History of the First 'Missing Link' and Its Discoverer* (Dordrecht: Kluwer Academic Publishers, 1989). Additional information is derived from the Dubois Archives, Nationaal Natuurhistorisch Museum, Leiden, The Netherlands.

37 " 'Dubois had the habit' " Theunissen, *Dubois,* p. 49, citing P. J. van der Feen and W. S. S. Van Bentham Jutting, "Antje Scheurder, Amsterdam, 15 november 1887–Amsterdam, 2 februari 1952," *Geologie en mijnbouw* 14 (1952): 122.

37 " 'So the question arises' " Theunissen, *Dubois,* p. 49, letter from Fürbringer to Dubois, October 2, 1886.

41–42 " 'Everything here has gone' " and " 'What's more, it was necessary' " Ibid., p. 40, letter from Dubois to F. A. Jentink, October 1889.

43 " 'A few cool windy days' " Eugène Dubois, note in Dubois Archives dated July 28, 1892.

43 "In August 1891" The account of the finding of *Pithecanthropus* and the replacement of the earlier name of *Anthropopithecus* is given in Theunissen, *Dubois.*

45 "On February 3, 1893, a local newspaper" *Homo erectus,* "Naar Aanleiding van Paleontologische Onderzoekingen op Java," *Bataviaasch Nieuwsblad,* Monday, February 6, 1803, no. 57. Clipping in Dubois Archives.

50 "He had a mirror over his front door" Frank Spencer, *Aleš Hrdlička, M.D. 1869–1943: A Chronicle of the Life and Work of an American Physical Anthropologist* (Ann Arbor: University Microfilms, 1979), pp. 418–21. Such mirrors can still be seen on houses in Haarlem today.

52 "He regarded the human ratio as perfect" Eugène Dubois, "The law of the necessary phylogenetic perfection of the Psychoencephalon,"

Proceedings of the Koninklijke Akademie van Wetenschappen te Amsterdam 31, no. 3 (1928): 304–14.

53 "Henry Fairfield Osborne, the self-important" Henry Fairfield Osborne, letter to L. Bok, secretary of the Royal Dutch Academy of Science, December 2, 1922; Henry Fairfield Osborne, letter to Dubois, April 2, 1924; both in Dubois Archives, Leiden.

53 "He acquiesced" See the extensive correspondence between Dubois and R. F. Damon and Company, London, that documents the negotiations for the casting program; see also correspondence between Dubois and Grafton Elliot Smith on this subject; all correspondence contained in the Dubois Archives, Leiden.

CHAPTER 3

54 "Davidson Black, a young Canadian physician" Information on Davidson Black's life and career is taken from: D. Hood, *Davidson Black, a Biography* (Toronto: University of Toronto Press, 1964); Jia Lanpo and Huang Weiwen, *The Story of Peking Man from Archaeology to Mystery,* trans. Yin Zhinqui (Beijing: Foreign Language Press, 1990); and Becky A. Sigmon and Jerome S. Cybulski, eds., *Homo erectus: Papers in Honor of Davidson Black* (Toronto: University of Toronto Press, 1981), especially the articles by Jerome S. Cybulski and Paul Gallina, "The works of Davidson Black," pp. 13–21, and H. L. Shapiro, "Davidson Black; An appreciation," pp. 21–26.

57 " 'jollied you, threatened you' " Spencer, *Hrdlička,* p. 775.

57–58 " 'a striking confirmation' " " 'new Tertiary man' " " 'The Chou Kou Tien discovery' " Lanpo and Weiwen, *Peking Man,* pp. 26–27.

58 " 'We have got' " Ibid., p. 49.

59 " 'a corking field man' " Ibid., p. 68.

59 " 'Found skullcap' " Ibid., p. 65.

60 " 'I made preparations' " Ibid., pp. 67–68.

61 "Dubois himself was never persuaded" Eugène Dubois, "Early Man in Java and *Pithecanthropus erectus,*" in *Early Man, as Depicted by Leading Authorities at the International Symposium, The Academy of Natural Sciences, Philadelphia, March 1937,* ed. G. G. MacCurdy, (Philadelphia: J. B. Lippincott Company, 1937), pp. 315–23.

62 "Excavation style" See Lanpo and Weiwen, *Peking Man,* pp. 81–92.

62 " 'Mechanize the Digging' " Ibid., p. 89.

63 "flowers on his grave" Shapiro, "Davidson Black; An appreciation," p. 24.

CHAPTER 4

64 " 'The past is a foreign country' " L. P. Hartley, *The Go-Between* (1953), prologue, in John Bartlett, comp., and Emily Morrison Beck, ed., *Familiar Quotations,* 15th ed. (Boston: Little Brown, 1980), p. 833.

65 "Black's successor, Franz Weidenreich" Information about his life and work taken from Loren C. Eiseley, "Franz Weidenreich, 1873–1948," *American Journal of Physical Anthropology* 7, no. 2 (1949): 241–53; W. K. Gregory, "Franz Weidenreich, 1873–1948," *American Anthropologist* 51, no. 1 (1949): 85–90; and Lanpo and Weiwen, *Peking Man.* The observation that Weidenreich's papers after leaving Germany were all written in English appears in the obituary by Gregory.

66–67 " 'The Cenozoic Laboratory' " Lanpo and Weiwen, *Peking Man,* p. 136.

67 "Despite Weidenreich's financial anxieties" Ibid., pp. 136–53.

67 "Weidenreich's newborn good luck" Different versions of the loss of the Peking Man fossils are given in Lanpo and Weiwen, *Peking Man,* and H. L. Shapiro, *Peking Man* (New York: Simon and Schuster, 1974).

71 " 'Where's Nelly?' " G. H. R. von Koenigswald, *Meeting Prehistoric Man* (London: Thames and Hudson, 1956), p. 51.

73 " 'However,' Weidenreich wrote" Franz Weidenreich, *Apes, Giants, and Man* (Chicago: University of Chicago Press, 1946), p. 3.

73 " 'prejudice and obstinate retention' " Franz Weidenreich, *The Skull of Sinanthropus pekinensis: A Comparative Study on a Primitive Hominid Skull,* Palaeontologica Sinica, New Series D no. 10 (1943): 214.

74 "To explain his hypothesis" Franz Weidenreich, "Facts and Speculations Concerning the Origin of *Homo sapiens,*" *American Journal of Physical Anthropology* 49, no. 2 (1947): 187–203.

75 "The mystery of the loss" Account taken from Shapiro, *Peking Man.*

77 "The newcomer was Christopher G. Janus" Account taken from Christopher G. Janus and William Brashler, *The Search for Peking Man* (New York: Macmillan, 1975).

79 "Finally, in 1981" Lanpo and Weiwen, *Peking Man,* p. 182, citing the *New York Times,* February 26, 1981.

CHAPTER 5

87 "Raymond Dart, an Australian anatomist" Information about Dart's life and career is taken from: Raymond A. Dart with Dennis Craig, *Adventures with the Missing Link* (Philadelphia: Institute Press, 1959); Frances Wheelhouse, *Raymond Arthur Dart: A Pictorial Profile* (Sydney: Transpareon Press, 1983); and an interview by the authors with G. W. H. Schepers, April 25, 1994.

88–89 " 'The incidents about tennis ball marks' " G. W. H. Schepers, letter to P.S., July 5, 1993, p. 3.

89 " 'One learnt absolutely' " Phillip V. Tobias, interview with P.S.; also Pat Shipman, "A journey towards human origins," *New Scientist,* October 19, 1991, pp. 45–47.

89 "In 1924, the only woman in the class" This account is taken from Dart and Craig, *Adventures,* p. 86.

92 " 'I was one of those' " Arthur Keith, *Autobiography* (London: Watts and Company, 1950), p. 480.

93 "nervous breakdown" Wheelhouse, *Dart,* p. 59; G. W. H. Schepers, interview with authors, April 25, 1994.

93 "For years, the Taung child sat, forgotten" G. W. H. Schepers, interview with authors, April 25, 1994.

93 " 'It seems likely that' " L. H. Wells, "One hundred years: Robert Broom, 30 November 1866–6 April 1951," *South African Journal of Science* 63, no. 9 (1967): 360.

94 "Not long afterward" " 'This was not inappropriate dress' " This account and quotations are taken from G. W. H. Schepers, interview with the authors, April 25, 1994. A slightly different account that downplays the role of Schepers and Harding le Riche is given in R. Broom, *Finding the Missing Link* (London: Watts and Company, 1950), pp. 39ff.

95 " 'Is this what you're after?' " Broom, *Finding,* p. 44.

95 " 'Competent Field Geologist' " Ibid., p. 62.

96 " 'Gen. Smuts was in America' " Ibid., p. 68.

96 " 'that there was no stratigraphy' " Ibid.

98 "it is telling" L. S. B. Leakey, *White African* (London: Holder and Stoughton, 1937), reprinted (Cambridge: Schenkman, 1966).

98 "If you want to understand" Information about Leakey's life and work are taken from Leakey, *White African;* also L. S. B. Leakey, *By the Evidence* (New York: Harcourt, Brace, Jovanovich, 1974); and Sonia Cole, *Leakey's Luck: The Life of Louis Leakey, 1903–1972* (London: Collins, 1975).

104 "Louis was initially suspicious" Cole, *Leakey's Luck,* p. 250ff.

104 "In 1961 Evernden, Louis, and Garniss Curtis" L. S. B. Leakey, J. F. Evernden, and Garniss Curtis, "Age of Bed I, Olduvai Gorge, Tanganyika," *Nature* 191 (1961): 478–79.

104 "The lower tuff" R. C. Walter et al., "Laser-fusion ^{40}Ar/^{39}Ar dating of Bed I, Olduvai Gorge, Tanzania," *Nature* 354 (1991): 145–49.

CHAPTER 6

107 "He coined two terms" John Napier, "The evolution of the hand," *Scientific American* 207 (1964): 308–12.

108 "The type specimen of *Homo habilis*" The original announcement of *Homo habilis* appears in L. S. B. Leakey, P. V. Tobias, and J. R. Napier, "A new species of the genus *Homo* from Olduvai Gorge," *Nature* 202 (1964): 7–9.

109–10 " 'The cranial capacity is very variable' " Ibid., p. 7.

110 " 'it is probable that the latter' " Ibid., p. 312.

111 "Louis, John, and Phillip were criticized" Shipman, "A journey," p. 47. For a fuller perspective, see P. V. Tobias, "The twenty years of rejection of *Homo habilis*" (paper delivered at the XIth Congress of U.I.S.P.P. in Mainz, West Germany, 1987).

125 "The specimen came to be known as OH 62" The primary description and announcement of this fossil is found in Donald C. Johanson et al., "New partial skeleton of *Homo habilis* from Olduvai Gorge, Tanzania," *Nature* 327 (1987): 205–9.

128 "Now the publication of . . . OH 62" R. E. Leakey et al., "A partial skeleton of a gracile hominid from the Upper Burgi Member of the Koobi Fora Formation, East Lake Turkana, Kenya," in *Hominidae: Proceedings of the 2nd International Congress of Human Paleontology, Turin,*

September 28–October 3, 1987, ed. Giacomo Giacobini (Milan: Jaca Books, 1989), pp. 167–74.

130 "a claim made a few years earlier by Randall Susman and Jack Stern" R. L. Susman and J. T. Stern, "Functional morphology of *Homo habilis,*" *Science* 217 (1982): 931–34.

132 "It was published in the *Journal of Human Evolution*" Sigrid Hartwig-Scherer and Robert Martin, "Was 'Lucy' more human than her 'child'? Observations on early hominid postcranial skeletons," *Journal of Human Evolution* 21 (1991): 439–49.

CHAPTER 7

139 "Robinson hypothesized that there was an ecological differ-ence" John Robinson, "Adaptive radiation in the australopithecines and the origin of man," in *African Ecology and Human Evolution,* ed. F. C. Howell and F. Bourlière (Chicago: Aldine, 1963), pp. 385–416.

140 "The genesis of the idea" C. Loring Brace, "The Fate of the 'Classic' Neanderthals: A Consideration of Hominid Catastrophism," *Current Anthropology* 5, no. 1 (1964): 3–43.

141 " 'Because of cultural adaptation' " Milford Wolpoff, "Compet-itive exclusion among Lower Pleistocene hominids: The single species hy-pothesis," *Man* 6, no. 4 (1971): 601.

141–2 "Milford believed John Robinson's differentiation" Milford Wolpoff, " 'Telanthropus' and the single species hypothesis," *American Anthropologist* 70, no. 3 (1968): 477–93.

145 "We wrote a draft" R. E. Leakey and A. Walker, "*Australo-pithecus, Homo erectus,* and the single species hypothesis," *Nature* 261 (1976): 572–74.

147 " 'the best-documented' " David Frayer, Milford Wolpoff, Alan Thorne, Fred Smith, and Geoffrey Pope, "Theories of modern human ori-gins: The paleontological test," *American Anthropologist* 95, no. 1 (1993): 18.

147 " 'The new data show' " Leakey and Walker, "*Australopithecus, Homo erectus,*" p. 574.

147, 149 "They soon found the famous partial skeleton known as Lucy" D. Johanson, T. D. White, and Y. Coppens, "A new species of the Genus *Australopithecus* (Primates: Hominidae) from the Pliocene of East-ern Africa," *Kirtlandia* 28 (1978): 1–14.

151 "Enrico had been studying hyraxes" Alan Walker, Heindrick Hoeck, and Linda Perez, "Microwear of mammalian teeth as an indicator of diet," *Science* 201 (1978): 908–10.

155 "My SEM work on *Australopithecus boisei*" Alan Walker, "Dietary hypotheses and human evolution," *Philosophical Transactions of the Royal Society, London* B 292 (1981): 57–64.

155 "He confirmed John Robinson's prediction" Fred Grine, "Trophic differences between 'gracile' and 'robust' australopithecines: A scanning electron microscope analysis of occlusal events," *South African Journal of Science* 77 (1981): 203–30.

155 "However, recent analyses of the chemical composition" A. Sillen, "Strontium-calcium (Sr/Ca) ratios of *Australopithecus robustus* and associated fauna from Swartkrans," *Journal of Human Evolution* 23 (1992): 495–516.

CHAPTER 8

158 "The selected pieces became a specimen numbered KNM-ER 1808" The description of 1808 and her diseased bones is given in Alan Walker, M. R. Zimmerman, and R. E. Leakey, "A possible case of hypervitaminosis A in *Homo erectus*," *Nature* 296 (1982): 248–50.

162 "*The Home of the Blizzard*" Details of Mawson's trip and trials are taken from Douglas Mawson, *The Home of the Blizzard* (London: Hodder and Stoughton, 1930), and Leonard Bickel, *Mawson's Will: The Greatest Survival Story Ever Written* (New York: Dorset Press, 1977).

164 "The excess of vitamin A" Bickel, *Mawson's Will*, p. 211.

165 " 'My God! Which one are you?' " Ibid.

169 "My best hope of understanding the dietary transition" Pat Shipman and Alan Walker, "The costs of becoming a predator," *Journal of Human Evolution* 18 (1989): 373–92.

CHAPTER 9

181 "Lucy is about 40 percent complete" Donald C. Johanson and Maitland Edey, *Lucy: The Beginnings of Mankind* (New York: Simon and Schuster, 1981); method of calculating completeness is explained in D. C. Johanson, letter to A.W., May 23, 1994.

183 " 'I wasn't going to pass that opportunity up' " The research of B. Holly Smith and quotations describing that work are taken from an interview with P.S. as well as from B. Holly Smith, "The physiological age of KNM-WT 15000," in *The Nariokotome* Homo erectus *Skeleton*, ed. Alan Walker and Richard Leakey (Cambridge: Harvard University Press, 1993), pp. 195–220.

189 "I had been working with Christopher Ruff" Christopher Ruff and Alan Walker, "Body size and body shape," in Walker and Leakey, *Nariokotome*, pp. 234–65.

193 "the invention of agriculture was distinctly detrimental" See, for example, M. N. Cohen and G. J. Armelagos, *Paleopathology at the Origins of Agriculture* (Orlando: Academic Press, 1984).

194 "For example, in 1883" Barbara Tuchman, *The Proud Tower: A Portrait of the World Before the War: 1890–1914* (New York: Bantam Books, 1966), p. 417, citing Simon Nowell-Smith, ed., *Edwardian England, 1901–1914,* (Oxford: Oxford University Press, 1964).

194 "I calculated that" Alan Walker, "Perspectives on the Nariokotome discovery," in Walker and Leakey, *Nariokotome*, pp. 411–32.

195 "Chris has concluded that U.S. blacks" Ruff and Walker, "Body size," p. 246.

195 "Like other mammals, humans adapt" Most of this work has been developed by Chris Ruff in Christopher Ruff, "Climate, body size, and body shape in hominid evolution," *Journal of Human Evolution* 21 (1991): 81–105, and Christopher Ruff, "Climatic Adaptation and Hominid Evolution: The Thermoregulatory Imperative," *Evolutionary Anthropology* 2, no. 2 (1993): 53–60.

197 "Bob Franciscus and Erik Trinkaus" Robert Franciscus and Erik Trinkaus, "Nasal morphology and the emergence of *Homo erectus*," *American Journal of Physical Anthropology* 75 (1988): 517–77.

199 "the boy was incredibly strong" Christopher Ruff et al., "Postcranial Robusticity in *Homo* I: Temporal Trends and Mechanical Interpretation," *American Journal of Physical Anthropology* 91 (1993): 21–53.

CHAPTER 10

207 "In all, Joan and I measured" Joan T. Richtsmeier and Alan Walker, "A morphometric study of facial growth," in Walker and Leakey, *Nariokotome*, pp. 411–32.

211 "Unlike a human of comparable size, the boy had only 880 cc" David Begun and Alan Walker, "The endocast," in Walker and Leakey, *Nariokotome,* pp. 326–58.

213 "A former colleague of mine . . . Marjorie Le May" Marjorie Le May, "Morphological cerebral asymmetries of modern man, fossil man and nonhuman primates," in *Origins and Evolution of Language and Speech,* ed. S. T. Harnard, H. D. Steklis, and J. Lancaster, Annals of the New York Academy of Sciences, 280 (1976): 349–66.

214 "The aspect of the boy's brain" Begun and Walker, "Endocast," pp. 326–58.

217 " 'unfeathered bipeds' " This is a paraphrase of a remark by Plato, *The Statesman,* in John Bartlett, comp., and Emily Morrison Beck, ed., *Familiar Quotations,* 15th ed. (Boston: Little Brown, 1980), p. 86.

218 "His narrow hips" See Alan Walker and Christopher Ruff, "The reconstruction of the pelvis," in Walker and Leakey, *Nariokotome,* pp. 221–33; also Alan Walker and Richard Leakey, "The Postcranial Bones," in Walker and Leakey, *Nariokotome,* pp. 95–160.

220 "In 1941 Adolf Portmann" The original work is A. Portmann, "Die Tragzeiten der Primaten und die Dauer der Schwangerschaft bein Menschen: ein Problem der vergleichenden Biologie," *Revue Suisse de Zoologie* 48 (1941): 511–18. An elaboration and expansion of those ideas in English was published in A. Portmann, *A Zoologist Looks at Humankind* (New York: Columbia University Press, 1990).

220 "In the early 1980s Robert Martin started reinvestigating" Robert Martin, "Relative brain size and basal metabolic rate in terrestrial vertebrates," *Nature* 293 (1981): 57–60; Robert Martin, "Human brain evolution in an ecological context," The 52nd Annual James Arthur Lecture on the Evolution of the Brain (American Museum of Natural History, New York, 1983).

223 "Because we had a complete braincase" The work that follows is taken from Begun and Walker, "Endocast," and Walker and Ruff, "Pelvis," both in Walker and Leakey, *Nariokotome.*

CHAPTER 11

230 "I decided to review" Pat Shipman and Alan Walker, "The costs of becoming a predator," *Journal of Human Evolution* 18 (1989): 373–92.

232–3 "To me, the most exciting paper" Leonid Gabunia and Antje Justus, "Une mandibule de l'homme fossile du Villefranchien Terminal de Dmanisi (Georgie Orientale)" (Paper delivered at the Fourth International Senckenberg Conference, "100 years of *Pithecanthropus*—the *Homo erectus* problem," December 2–6, 1991). L. Gabunia and A. Vekua, "A Plio-Pleistocene hominid from Dmanisi, East Georgia, Caucasus," *Nature* 373 (1995): 509–12.

234 "These dates suggested that *Homo erectus* reached Java" C. Swisher III et al., "Age of the earliest known hominids in Java, Indonesia," *Science* 263 (1994): 1118–21.

235 "The first serious attempt to date the Javan sites" N. Watanabe and D. Kadar, eds., *Quaternary Geology of the Hominid Fossil-bearing Formations in Java* (Geological Research and Development Centre special publication 4, Bandung, Indonesia, 1986); also, Masayuki Hyodo et al. "Magnetostratigraphy of Hominid Fossil Bearing Formations in Sangiran and Mojokerto, Java," *Anthropological Science* 101, no. 2 (1993): 157–86.

236 " 'really human' " " 'nothing in common' " Eugène Dubois, "Early Man in Java," in MacCurdy, *Early Man,* p. 315.

237 "Various other publications by von K" For a summary of the statements about the Mojokerto site, see John de Vos, "Faunal stratigraphy and correlation of the Indonesian hominid sites," in *Ancestors: The Hard Evidence,* ed. E. Delson (New York: Alan R. Liss, 1984), pp. 215–20, esp. pp. 218–19.

237 "Swisher thought" Carl Swisher III, interview with P.S.

242 "Lyman Jellema and Bruce Latimer . . . joined me in analyzing the boy's rib cage" Lyman M. Jellema, Bruce Latimer, and Alan Walker, "The rib cage," in Walker and Leakey, *Nariokotome,* pp. 294–325.

243 "Leslie Aiello . . . and Peter Wheeler" Leslie Aiello and Peter Wheeler, "Brains and guts in human and primate evolution: The expensive organ hypothesis," *Current Anthropology* 36, no. 2 (1995): 199–221.

CHAPTER 12

250 "Carol and Bruce wanted to discover" Bruce Latimer and Carol Ward, "The thoracic and lumbar vertebrae," in Walker and Leakey, *Nariokotome,* pp. 266–93.

252 "a young Dutch anatomist, Fred Spoor" C. F. Spoor, "The Comparative Morphology and Phylogeny of the Human Bony Labyrinth," Ph.D. dissertation (Utrecht: University of Utrecht, 1993). See also C. F. Spoor, F. Zonneveld, and B. Wood, "Early hominid labyrinthine morphology and its possible implications for the evolution of human bipedal locomotion," *Nature* 369 (1994): 645–48.

258 " 'It's very difficult to interpret' " " 'The only thing' " " 'Either this specimen is not' " Fred Spoor, interview with P.S. See also Spoor, *Comparative Morphology*, pp. 109–35.

CHAPTER 13

263 "Ann's previous studies" Ann MacLarnon, "Size relationships and the spinal cord and associated skeleton in primates," Ph.D. dissertation (London: University of London, 1987), and Ann MacLarnon, "The vertebral canal," in Walker and Leakey, *Nariokotome*, pp. 359–90.

265 " 'I gave a seminar' " Ann MacLarnon, letter to A.W.

267 "One of the first things Raichle's group decided to investigate" See M. Raichle et al., "Positron Emission Tomography studies of the cortical anatomy of single-word processing," *Nature* 331 (1988): 585–89; M. Raichle et al., "Positron Emission Tomography studies of the processing of single words," *Journal of Cognitive Science* 1, no. 2 (1989): 153–70.

269 " 'that marvellous endowment' " T. H. Huxley, *Man's Place in Nature and Other Anthropological Essays* (New York: D. Appleton and Co., 1900), p. 155.

271 "In 1868, when Ernst Haeckel" See Haeckel, *History of Creation*.

272 "In a recent study of deaf children" L. A. Petitto and P. F. Marentette, "Babbling in the manual mode: Evidence for the ontogeny of language," *Science* 251 (1991): 1493–96.

272 " 'language instinct' " One of the most cogent discussions of human language is given by Steven Pinker, *The Language Instinct: How the Mind Creates Language* (New York: Morrow, 1994).

272 "A well-documented example" The primary reference is S. Curtiss, *Genie: A Psycholinguistic Study of a Modern-Day "Wild Child"* (New York: Academic Press, 1977). Genie's case is also discussed in Pinker, *Language Instinct,* and Derek Bickerton, *Language and Species* (Chicago: University of Chicago Press, 1990).

274 "Animal language works" See, for example, Sue Savage-Rumbaugh, *Ape Language: From Conditioned Response to Symbol* (New York: Columbia University Press, 1986); Sue Savage-Rumbaugh and Roger Lewin, *Kanzi: The Ape at the Brink of the Human Mind* (New York: John Wiley, 1994); or Dorothy L. Cheney and Robert M. Seyfarth, *How Monkeys See the World* (Chicago: University of Chicago Press, 1990.)

274 "True or full language" Bickerton, *Language and Species.*

CHAPTER 14

280 "The first, groundbreaking attempt" P. Lieberman and E. Crelin, "On the speech of neanderthal man," *Linguistic Inquiry* 2 (1971): 203–22. Lieberman's more recent thoughts on the subject are presented in Phillip Lieberman, *Biology and Evolution of Language* (Cambridge: Harvard University Press, 1984).

281 "But an anatomist/anthropologist of the next generation" Jeffrey Laitman, "Evolution of the hominid upper respiratory tract: The fossil evidence," in *Hominid Evolution: Past, Present and Future,* ed. P. V. Tobias (New York: Alan Liss, 1985), pp. 281–86.

282 "The 'language written in stone' idea" R. Holloway, "Culture: A *human* domain," *Current Anthropology* 10 (1969): 395–407.

283 "Two anthropologists at the University of New England" William Noble and Iain Davidson, "The evolutionary emergence of modern human behavior: Language and its archeology," *Man* 26 (1991): 223–54.

285 "There are hints that symbolic behavior" For another review of the evidence for language, and a conclusion different from Noble and Davidson, see L. A. Schepartz, "Language and Modern Human Origins," *Yearbook of Physical Anthropology* 36 (1993): 91–126.

286 "studies conducted by Olga Soffer" Olga Soffer, "Before Beringia: Late Pleistocene bio-social transformation and the colonization of northern Eurasia," in *Chronostratigraphy of the Paleolithic in North, Central, East Asia and America* (Novosibirsk: Academy of Sciences of the USSR, 1990). See also Erik Trinkaus, ed., *The Emergence of Modern Humans: Biocultural Adaptations in the Later Pleistocene* (Cambridge: Cambridge University Press, 1989), and Paul Mellars, ed., *The Emergence of Modern Humans: An Archaeological Perspective* (Ithaca: Cornell University Press, 1990), and papers therein.

287 "As Randall White . . . has observed" Randall White, "Rethinking the Middle/Upper Paleolithic Transition," *Current Anthropology* 23 (1982): 169–92; Randall White, "Thoughts on social relationships and language in hominid evolution," *Journal of Social and Personal Relationships* 2 (1985): 95–115; and Randall White, "Toward an understanding of the earliest body ornaments," in E. Trinkaus, *Emergence,* pp. 211–31.

288 "Margaret Conkey, a Berkeley anthropologist" Margaret Conkey, "Style and information in cultural evolution; Toward a predictive model for the Paleolithic," in *Social Archaeology,* ed. C. L. Redman et al. (New York: Academic Press, 1978), pp. 61–85; M. Conkey, "The identification of prehistoric hunter-gatherer aggregation sites: The case of Altamira," *Current Anthropology* 21 (1980): 609–30; M. Conkey, "On the origins of Paleolithic art; A review and some critical thoughts," in *The Mousterian Legacy: Human Biocultural Change in the upper Pleistocene,* ed. E. Trinkaus (Oxford: British Archaeological Reports S164, 1983), pp. 201–27.

288–9 "Archeologist Sandra Bowdler . . . sees the colonization of Greater Australia" Sandra Bowdler, "Peopling Australasia: the 'Coastal Colonization' Hypothesis Re-examined," in *The Emergence of Modern Humans,* ed. Paul Mellars (Ithaca: Cornell University Press, 1990), pp. 327–43.

289–90 "An entirely different type of evidence" Luigi Cavalli-Sforza, "Genes, Peoples, and Languages," *Scientific American,* November 1991, pp. 104–10; Luigi Cavalli-Sforza et al., "Reconstruction of human evolution; Bringing together genetic, archaeological and linguistic data," *Proceedings of the National Academy of Sciences* 85 (1988): 6002–6; Luigi Cavalli-Sforza, Eric Minch, and J. L. Mountain, "Coevolution of genes and languages revisited," *Proceedings of the National Academy of Sciences* 89 (1992): 5620–24.

291 "Cavalli-Sforza gives an example" See Cavalli-Sforza, "Genes, Peoples," p. 110.

292 " 'if they were biologically' " " 'In our society' " Cavalli-Sforza et al., "Reconstruction," p. 6006.

BIBLIOGRAPHY

ARCHIVAL MATERIALS

The following items are contained in the Dubois Archives, Nationaal
Natuurhistorisch Museum, Leiden, The Netherlands:

Anonymous (under the pseudonym *Homo erectus*), "Naar Aanleiding van
   Paleontologische Onderzoekingen op Java," *Bataviaasch Nieuws-
   blad,* Monday, February 6, 1803, no. 57.
Dubois, Eugène. Correspondence with R. F. Damon and Company, Lon-
   don.
————. Correspondence with G. Elliot Smith, London.
————. Note, dated July 28, 1892.
Osborne, Henry Fairfield. Letter to L. Bok, secretary of the Royal Dutch
   Academy of Science, December 2, 1922.
————. Letter to Eugène Dubois, April 2, 1924.

BOOKS AND ARTICLES

Aiello, Leslie, and Wheeler, Peter. "Brains and guts in human and primate evolution: The expensive organ hypothesis." *Current Anthropology* 36, no. 2 (1995): 199–221.

Begun, David, and Walker, Alan. "The endocast." In *The Nariokotome Homo erectus Skeleton,* edited by Alan Walker and Richard Leakey, pp. 326–58. Cambridge: Harvard University Press, 1993.

Bickel, Leonard. *Mawson's Will: The Greatest Survival Story Ever Written.* New York: Dorset Press, 1977.

Bickerton, Derek. *Language and Species.* Chicago: University of Chicago Press, 1990.

Bowdler, Sandra. "Peopling Australasia: the 'Coastal Colonization' Hypothesis Re-examined." In *The Emergence of Modern Humans,* edited by Paul Mellars, pp. 327–43. Ithaca: Cornell University Press, 1990.

Brace, C. Loring. "The Fate of the 'Classic' Neanderthals: A Consideration of Hominid Catastrophism." *Current Anthropology* 5, no. 1 (1964): 3–43.

Broom, Robert. *Finding the Missing Link.* London: Watts and Co., 1950.

Cavalli-Sforza, Luigi. "Genes, Peoples, and Languages." *Scientific American,* November 1991, pp. 104–10.

Cavalli-Sforza, Luigi; Minch, Eric; and Mountain, J. L. "Coevolution of genes and languages revisited." *Proceedings of the National Academy of Sciences* 89 (1992): 5620–24.

Cavalli-Sforza, Luigi; Piazza, Alberto; Menozzi, Paolo; and Mountain, Joanna. "Reconstruction of human evolution; Bringing together genetic, archaeological and linguistic data." *Proceedings of the National Academy of Sciences* 85 (1988): 6002–6.

Cheney Dorothy L., and Seyfarth, Robert M. *How Monkeys See the World.* Chicago: University of Chicago Press, 1990.

Cohen, M. N., and Armelagos, G. J. *Paleopathology at the Origins of Agriculture.* Orlando: Academic Press, 1984.

Cole, Sonia. *Leakey's Luck: The Life of Louis Leakey, 1903–1972.* London: Collins, 1975.

Conkey, Margaret. "Style and information in cultural evolution; Toward a predictive model for the Paleolithic." In *Social Archaeology,* edited by C. L. Redman, M. J. Berman, W. T. Longhorne Jr., N. M. Versagg, and J. C. Wanser, pp. 61–85. New York: Academic Press, 1978.

————. "The identification of prehistoric hunter-gatherer aggregation sites: The case of Altamira." *Current Anthropology* 21 (1980): 609–30.

————. "On the origins of Paleolithic art; A review and some critical thoughts." In *The Mousterian Legacy; Human Biocultural Change in the upper Pleistocene,* edited by E. Trinkaus, pp. 201–27. Oxford: British Archaeological Reports S164, 1983.

Curtiss, S. *Genie: A Psycholinguistic Study of a Modern-Day "Wild Child."* New York: Academic Press, 1977.

Cybulski, Jerome S., and Gallina, Paul. "The works of Davidson Black." In *Homo erectus: Papers in Honor of Davidson Black,* edited by Becky A. Sigmon and Jerome S. Cybulski, pp. 13–21. Toronto: University of Toronto Press, 1981.

Dart, Raymond A., with Craig, Dennis. *Adventures with the Missing Link.* Philadelphia: Institutes Press, 1959.

Darwin, F., and Seward, A. C., eds. *More Letters of Charles Darwin.* London: Murray, 1903.

de Vos, John. "Faunal stratigraphy and correlation of the Indonesian hominid sites." In *Ancestors: The Hard Evidence,* edited by E. Delson, pp. 215–20. New York: Alan R. Liss, 1984.

Dubois, Eugène. "Early Man in Java and *Pithecanthropus erectus.*" In *Early Man, as Depicted by Leading Authorities at the International Symposium, The Academy of Natural Sciences, Philadelphia, March 1937,* edited by G. G. MacCurdy, pp. 315–23. Philadelphia: J. B. Lippincott Company, 1937.

————. "The law of the necessary phylogenetic perfection of the Psychoencephalon." *Proceedings of the Koninklijke Akademie van Wetenschappen te Amsterdam* 31, no. 3 (1928): 304–14.

Eiseley, Loren C. "Franz Weidenreich, 1873–1948." *American Journal of Physical Anthropology* 7, no. 2 (1949): 241–53.

Franciscus, Robert, and Trinkaus, Erik. "Nasal morphology and the emergence of *Homo erectus.*" *American Journal of Physical Anthropology* 75 (1988): 517–77.

Frayer, David; Wolpoff, Milford; Thorne, Alan; Smith, Fred; and Pope, Geoffrey. "Theories of modern human origins: The paleontological test." *American Anthropologist* 95, no. 1 (1993): 14–50.

Gabunia, Leonid, and Justus, Antje. "Une mandibule de l'homme fossile du Villefranchien Terminal de Dmanisi (Georgie Orientale)." Paper delivered at the Fourth International Senckenberg Conference, "100 years of *Pithecanthropus*—the *Homo erectus* problem," December 2–6, 1991. Gabunia, L., and Vekua, A. "A Plio-Pleistocene hominid from Dmanisi, East Georgia, Caucasus." *Nature* 373 (1995): 509–12.

Gregory, W. K. "Franz Weidenreich, 1873–1948." *American Anthropologist* 51, no. 1 (1949): 85–90.

Grine, Fred. "Trophic differences between 'gracile' and 'robust' australopithecines: A scanning electron microscope analysis of occlusal events." *South African Journal of Science* 77 (1981): 203–30.

Haeckel, Ernst. *The History of Creation, or the Development of the Earth and Its Inhabitants by the Action of Natural Causes. A Popular Exposition of the Doctrine of Evolution in General, and That of Darwin, Goethe, and Lamarck in Particular.* Translated by E. Ray Lankester. New York: D. Appleton, 1868.

Hartwig-Scherer, Sigrid, and Martin, Robert. "Was 'Lucy' more human than her 'child'? Observations on early hominid postcranial skeletons." *Journal of Human Evolution* 21 (1991): 439–49.

Holloway, Ralph. "Culture: a *human* domain." *Current Anthropology* 10 (1969): 395–407.

Hood, D. *Davidson Black, a Biography.* Toronto: University of Toronto Press, 1964.

Huxley, T. H. *Man's Place in Nature and Other Anthropological Essays.* New York: D. Appleton and Co., 1900.

Hyodo, Masayuki; Watanabe, Naotune; Sunata, Wahyu; Susanto, Eko Edi; and Wahyong, Hendro. "Magnetostratigraphy of Hominid Fossil Bearing Formations in Sangiran and Mojokerto, Java." *Anthropological Science* 101, no. 2 (1993): 157–86.

Janus, Christopher G., and Brashler, William. *The Search for Peking Man.* New York: Macmillan, 1975.

Jellema, Lyman M.; Latimer, Bruce; and Walker, Alan. "The rib cage." In *The Nariokotome* Homo erectus *Skeleton,* edited by Alan Walker and Richard Leakey, pp. 266–93. Cambridge: Harvard University Press, 1993.

Johanson, Donald C., and Edey, Maitland. *Lucy: The Beginnings of Mankind.* New York: Simon and Schuster, 1981.

Johanson, Donald C.; White, T. D.; and Coppens, Y. "A new species of the Genus *Australopithecus* (Primates: Hominidae) from the Pliocene of Eastern Africa." *Kirtlandia* 28 (1978): 1–14.

Johanson, Donald C.; Masao, Fidelis T.; Eck, Gerald G.; White, Tim D.; Walter, Robert C.; Kimbel, William H.; Asfaw, Berhane; Manega, Paul; Ndessokia, Prosper; and Suwa, Gen. "New partial skeleton of *Homo habilis* from Olduvai Gorge, Tanzania." *Nature* 327 (1987): 205–9.

Keith, Arthur. *Autobiography*. London: Watts and Company, 1950.

Kimeu, Kamoya. "Adventures in the Bone Trade." *Science 86* 7 (1986): 39–41.

Laitman, Jeffrey. "Evolution of the hominid upper respiratory tract: The fossil evidence." In *Hominid Evolution: Past, Present and Future*, edited by P. V. Tobias, pp. 281–86. New York: Alan Liss, 1985.

Lanpo, Jia, and Weiwen, Huang. *The Story of Peking Man from Archaeology to Mystery*. Translated by Yin Zhinqui. Beijing: Foreign Language Press, 1990.

Latimer, Bruce, and Ward, Carol. "The thoracic and lumbar vertebrae." In *The Nariokotome* Homo erectus *Skeleton,* edited by Alan Walker and Richard Leakey, pp. 266–93. Cambridge: Harvard University Press, 1993.

Leakey, L. S. B. *By the Evidence*. New York: Harcourt, Brace, Jovanovich, 1974.

———. *White African*. London: Holder and Stoughton, 1937. Reprinted. Cambridge: Schenkman, 1966.

Leakey, L. S. B.; Evernden, J. F.; and Curtis, Garniss. "Age of Bed I, Olduvai Gorge, Tanganyika." *Nature* 191 (1961): 478–79.

Leakey, L. S. B.; Tobias, P. V.; and Napier, J. R. "A new species of the genus *Homo* from Olduvai Gorge." *Nature* 202 (1964): 308–12.

Leakey, R. E., and Walker, A. *"Australopithecus, Homo erectus,* and the single species hypothesis." *Nature* 261 (1976): 572–74.

Leakey, R. E.; Walker, Alan; Ward, C. V.; and Grausz, H. M. "A partial skeleton of a gracile hominid from the Upper Burgi Member of the Koobi Fora Formation, East Lake Turkana, Kenya." In *Hominidae: Proceedings of the 2nd International Congress of Human Paleontology, Turin, September 28–October 3, 1987,* edited by Giacomo Giacobini, pp. 167–74. Milan: Jaca Books, 1989.

Le May, Marjorie. "Morphological cerebral asymmetries of modern man, fossil man and nonhuman primates." In *Origins and Evolution of Language and Speech,* edited by S. T. Harnard, H. D. Steklis, and J. Lancaster, pp. 349–66. Annals of the New York Academy of Sciences, 280 (1976).

Lieberman, P. *Biology and Evolution of Language.* Cambridge: Harvard University Press, 1984.

Lieberman, P., and Crelin, E. "On the speech of neanderthal man." *Linguistic Inquiry* 2 (1971): 203–22.

Litchfield, Henrietta, ed. *Emma Darwin: A Century of Family Letters, 1792–1896.* London: Murray, 1915.

McCown, T. D., and Kennedy, K. A. R., eds. *Climbing Man's Family Tree: A Collection of Major Writings on Human Phylogeny, 1699 to 1971.* Englewood Cliffs, N. J.: Prentice-Hall, 1972.

MacLarnon, Ann. "Size relationships and the spinal cord and associated skeleton in primates." Ph.D. dissertation, University of London, 1987.

———. "The vertebral canal." In *The Nariokotome* Homo erectus *Skeleton,* edited by Alan Walker and Richard Leakey, pp. 359–90. Cambridge: Harvard University Press, 1993.

Martin, Robert. "Relative brain size and basal metabolic rate in terrestrial vertebrates." *Nature* 293 (1981): 57–60.

———. "Human brain evolution in an ecological context." The 52nd Annual James Arthur Lecture on the Evolution of the Brain (American Museum of Natural History, New York, 1983).

Mawson, Douglas. *The Home of the Blizzard.* London: Hodder and Stoughton, 1930.

Mellars, Paul, ed. *The Emergence of Modern Humans: An Archaeological Perspective.* Ithaca: Cornell University Press, 1990.

Napier, John. "The evolution of the hand." *Scientific American* 207 (1964): 308–12.

Noble, William, and Davidson, Iain. "The evolutionary emergence of modern human behavior: Language and its archeology." *Man* 26 (1991): 223–54.

Owen, Richard. "Of the Anthropoid Apes and their relations to Man." *Proceedings of the Royal Institute of Great Britain 1854–1858* 3 (1855): 26–41.

Petersen, S. E.; Fox, P. T.; Posner, M. I.; Mintun, M.; and Raichle, M. "Positron Emission Tomography studies of the cortical anatomy of single-word processing." *Nature* 331 (1988): 585–89.

———. "Positron Emission Tomography studies of the processing of single words." *Journal of Cognitive Science* 1, no. 2 (1989): 153–70.

Petitto, L. A., and Marentette, P. F. "Babbling in the manual mode: Evidence for the ontogeny of language." *Science* 251 (1991): 1493–96.

Pinker, Steven. *The Language Instinct: How the Mind Creates Language.* New York: Morrow, 1994.

Portmann, Adolf. "Die Tragzeiten der Primaten und die Dauer der Schwangerschaft bein Menschen: ein Problem der vergleichenden Biologie." *Revue Suisse de Zoologie* 48 (1941): 511–18.

———. *A Zoologist Looks at Humankind.* New York: Columbia University Press, 1990.

Richtsmeier, Joan T., and Walker, Alan. "A morphometric study of facial growth." In *The Nariokotome* Homo erectus *Skeleton,* edited by Alan Walker and Richard Leakey, pp. 411–32. Cambridge: Harvard University Press, 1993.

Robinson, John. "Adaptive radiation in the australopithecines and the origin of man." In *African Ecology and Human Evolution,* edited by F. C. Howell and F. Bourlière, pp. 385–416. Chicago: Aldine, 1963.

Ruff, Christopher. "Climate, body size, and body shape in hominid evolution." *Journal of Human Evolution* 21 (1991): 81–105.

———. "Climatic Adaptation and Hominid Evolution: The Thermoregulatory Imperative." *Evolutionary Anthropology* 2, no. 2 (1993): 53–60.

Ruff, Christopher, and Walker, Alan. "Body size and body shape." In *The Nariokotome* Homo erectus *Skeleton,* edited by Alan Walker and Richard Leakey, pp. 234–65. Cambridge: Harvard University Press, 1993.

Ruff, Christopher; Trinkaus, Erik; Walker, Alan; and Larsen, Clark Spencer. "Postcranial Robusticity in *Homo* I: Temporal Trends and Mechanical Interpretation." *American Journal of Physical Anthropology* 91 (1993): 21–53.

Savage-Rumbaugh, Sue. *Ape Language: From Conditioned Response to Symbol.* New York: Columbia University Press, 1986.

Savage-Rumbaugh, Sue, and Lewin, Roger. *Kanzi: The Ape at the Brink of the Human Mind.* New York: John Wiley, 1994.

Schepartz, L. A. "Language and Modern Human Origins." *Yearbook of Physical Anthropology* 36 (1993): 91–126.

Shapiro, H. L. "Davidson Black; An appreciation." In *Homo erectus: Papers in Honor of Davidson Black,* edited by Becky A. Sigmon and Jerome S. Cybulski, pp. 21–26. Toronto: University of Toronto Press, 1981.

————. *Peking Man.* New York: Simon and Schuster, 1974.

Shipman, Pat. "A journey towards human origins." *New Scientist*, October 19, 1991, pp. 45–47.

Shipman, Pat, and Walker, Alan. "The costs of becoming a predator." *Journal of Human Evolution* 18 (1989): 373–92.

Sigmon, Becky A., and Cybulski, Jerome S., eds. *Homo erectus: Papers in Honor of Davidson Black.* Toronto: University of Toronto Press, 1981.

Sillen, Andrew. "Strontium-calcium (Sr/Ca) ratios of *Australopithecus robustus* and associated fauna from Swartkrans." *Journal of Human Evolution* 23 (1992): 495–516.

Smith, B. Holly. "The physiological age of KNM-WT 15000." In *The Nariokotome* Homo erectus *Skeleton,* edited by Alan Walker and Richard Leakey, pp. 195–220. Cambridge: Harvard University Press, 1993.

Soffer, Olga. "Before Beringia: Late Pleistocene bio-social transformation and the colonization of northern Eurasia." In *Chronostratigraphy of the Paleolithic in North, Central, East Asia and America.* Novosibirsk: Academy of Sciences of the USSR, 1990.

Spencer, Frank. *Aleš Hrdlička, M.D. 1869–1943: A Chronicle of the Life and Work of an American Physical Anthropologist.* Ann Arbor: University Microfilms, 1979.

Spoor, C. F. "The Comparative Morphology and Phylogeny of the Human Bony Labyrinth." Ph.D. dissertation, University of Utrecht, 1993.

Spoor, C. F.; Zonneveld, F.; and Wood, B. "Early hominid labyrinthine morphology and its possible implications for the evolution of human bipedal locomotion." *Nature* 369 (1994): 645–48.

Susman, R. L., and Stern, J. T. "Functional morphology of *Homo habilis.*" *Science* 217 (1982): 931–34.

Swisher, Carl, III; Curtis, G. H.; Jacob, T.; Getty, A. G.; and Widlasmoro, A. Suprijo. "Age of the earliest known hominids in Java, Indonesia." *Science* 263 (1994): 1118–21.

Theunissen, Bert. *Eugène Dubois and the Ape-Man from Java: The History of the First 'Missing Link' and Its Discoverer.* Dordrecht: Kluwer Academic Publishers, 1989.

Tobias, P. V. "The twenty years of rejection of *Homo habilis.*" Paper delivered at the XIth Congress of U.I.S.P.P. in Mainz, West Germany, 1987.

Trinkaus, Erik, ed. *The Emergence of Modern Humans: Biocultural Adaptations in the Later Pleistocene.* Cambridge: Cambridge University Press, 1989.

Tuchman, Barbara. *The Proud Tower: A Portrait of the World Before the War: 1890–1914.* New York: Bantam Books, 1966.

van der Feen, P. J., and Jutting, W. S. S. Van Bentham. "Antje Scheurder, Amsterdam, 15 november 1887–Amsterdam, 2 februari 1952." *Geologie en mijnbouw* 14 (1952): 121–25.

von Koenigswald, G. H. R. *Meeting Prehistoric Man.* London: Thames and Hudson, 1956.

Walker, Alan. "Dietary hypotheses and human evolution." *Philosophical Transactions of the Royal Society, London* B 292 (1981): 57–64.

———. "Perspectives on the Nariokotome discovery." In *The Nariokotome* Homo erectus *Skeleton,* edited by Alan Walker and Richard Leakey, pp. 411–32. Cambridge: Harvard University Press, 1993.

Walker, Alan, and Leakey, R. E. "The hominids of East Turkana." *Scientific American* 239, no. 2 (1978): 54–66.

Walker, Alan, and Leakey, Richard. "The Postcranial Bones." In *The Nariokotome* Homo erectus *Skeleton,* edited by Alan Walker and Richard Leakey, pp. 95–160. Cambridge: Harvard University Press, 1993.

Walker, Alan, and Ruff, Christopher. "The reconstruction of the pelvis." In *The Nariokotome* Homo erectus *Skeleton,* edited by Alan Walker and Richard Leakey, pp. 221–33. Cambridge: Harvard University Press, 1993.

Walker, Alan; Hoeck, Heindrick; and Perez, Linda. "Microwear of mammalian teeth as an indicator of diet." *Science* 201 (1978): 908–10.

Walker, Alan; Zimmerman, M. R.; and Leakey, R. E. "A possible case of hypervitaminosis A in *Homo erectus.*" *Nature* 296 (1982): 248–50.

Walter, R. C.; Manega, P. C.; Hay, R. L.; Drake, R. E.; and Curtis, Garniss. "Laser-fusion ^{40}Ar/^{39}Ar dating of Bed I, Olduvai Gorge, Tanzania." *Nature* 354 (1991): 145–49.

Watanabe, N., and Kadar, D., eds. *Quaternary Geology of the Hominid Fossil-bearing Formations in Java.* Geological Research and Development Centre special publication 4, Bandung, Indonesia, 1986.

Weidenreich, Franz. *Apes, Giants, and Man.* Chicago: University of Chicago Press, 1946.

———. *The Skull of Sinanthropus pekinensis: A Comparative Study on a Primitive Hominid Skull.* Palaeontologica Sinica, New Series D No. 10 (1943): 1–485.

———. "Facts and Speculations Concerning the Origin of *Homo sapiens.*" *American Journal of Physical Anthropology* 49, no. 2 (1947): 187–203.

Wells, L. H. "One hundred years: Robert Broom, 30 November 1866–6 April 1951." *South African Journal of Science* 63, no. 9 (1967): 357–66.

Wheelhouse, Frances. *Raymond Arthur Dart: A Pictorial Profile.* Sydney: Transpareon Press, 1983.

White, Randall. "Rethinking the Middle/Upper Paleolithic Transition." *Current Anthropology* 23 (1982): 169–92.

———. "Thoughts on social relationships and language in hominid evolution." *Journal of Social and Personal Relationships* 2 (1985): 95–115.

———. "Toward an understanding of the earliest body ornaments." In *The Emergence of Modern Humans: Biocultural Adaptations in the Later Pleistocene,* edited by E. Trinkaus, pp. 211–31. Cambridge: Cambridge University Press, 1989.

Wolpoff, Milford. "Competitive exclusion among Lower Pleistocene hominids: The single species hypothesis." *Man* 6, no. 4 (1971): 601–14.

———. " 'Telanthropus' and the single species hypothesis." *American Anthropologist* 70, no. 3 (1968): 477–93.

INDEX

Italicized page numbers refer to illustrations.

ILLUSTRATION CREDITS

Illustrations not otherwise credited were drawn or photographed by
Alan Walker.

Page 138: A. Walker and R. E. Leakey, "The hominids of East
Turkana," *Scientific American* 239, no. 2 (1978): 54–66. Copyright ©
1978 by Scientific American, Inc. All rights reserved.

148: Photo by R. E. Leakey.

189: After Smith in Walker and Leakey, *Nariokotome,* figure 9.3.
Reprinted by permission of the publishers from *The Nariokotome*
homo erectus *Skeleton,* edited by Alan Walker and Richard Leakey
(Cambridge, Mass.: Harvard University Press). Copyright © 1993 by
Alan Walker and Richard Leakey.

206: Drawing by Elaine Kasmer from Richtsmeier in Walker and
Leakey, *Nariokotome,* figure 16.1a. Reprinted by permission of the
publishers from *The Nariokotome* homo erectus *Skeleton,* edited by
Alan Walker and Richard Leakey (Cambridge, Mass.: Harvard
University Press). Copyright © 1993 by Alan Walker and Richard
Leakey.

224: After Martin, "Human brain evolution," 1983.

244: Drawing by Elaine Kasmer from Ruff, "Climatic Adaptation," figure 2.

Insert figures 3, 4. Photo by Virginia Morell.

Insert figure 6. Photo by Michael McRae.

Insert figure 8. Courtesy of John de Vos, Dubois Archives, copyright Nationaal Natuurhistorisch Museum, Leiden, The Netherlands.

Insert figure 10. Unable to locate copyright holder.

Insert figure 11. (a) Photo by R. E. Leakey. (b) Drawing by Christine Young © 1980 Alan Walker.

Insert figure 14. W. W. Howells, "The distribution of man," *Scientific American* 203 (1960): 112–27. Copyright © 1960 by Scientific American, Inc. All rights reserved.

Insert figure 17. Drawing after Michael Posner and Marcus Raichle, *Images of the Mind* (New York: Scientific American Library, 1994), p. 115.

Insert figure 18. Drawing after Cavalli-Sforza et al., "Coevolution."

A Note About the Authors

ALAN WALKER is professor of anthropology and biology at the Pennsylvania State University and is a veteran of dozens of field trips searching for fossils in Africa. Among his recent honors are a Guggenheim Fellowship, a MacArthur Foundation Fellowship, and a year spent as a Phi Beta Kappa Visiting Scholar. He is the author of a technical book on Nariokotome boy (*The Nariokotome Homo erectus Skeleton*) and has written for *Scientific American*.

PAT SHIPMAN is adjunct professor of anthropology at the Pennsylvania State University and has researched human origins on her own and collaboratively with Alan Walker, her husband. She is the author of *The Evolution of Racism* and (with Erik Trinkaus) *The Neandertals*. Her articles on science and evolution have appeared in *New Scientist*, *The Sciences*, and *Natural History*. She and her husband live in State College, Pennsylvania.

A Note on the Type

This book was set on the Linotype in Granjon, a type named in compliment to Robert Granjon, a type cutter and printer active in Antwerp, Lyons, Rome, and Paris, from 1523 to 1590. Granjon, the boldest and most original designer of his time, was one of the first to practice the trade of type founder apart from that of printer.

Linotype Granjon was designed by George W. Jones, who based his drawings on a face used by Claude Garamond (c. 1480–1561) in his beautiful French books. Granjon more closely resembles Garamond's own type than does any of the various modern faces that bear his name.

Composed by North Market Street Graphics,
Lancaster, Pennsylvania
Printed and bound by R. R. Donnelley & Sons,
Harrisonburg, Virginia
Designed by Brooke Zimmer